# THE CELL

ENCYCLOPAEDIA OF CELL BIOLOGY - I

# THE CELL

*By*

**Dr. M.Prakash**

*Dept. of Zoology*
*M.M.H. Post Graduate College*
*Ghaziabad*
*(U.P.)*

First Published – 2010

Reprinted – 2025

ISBN: 978-81-8356-454-0 (Set)
978-81-8356-559-2



**The Cell**

*Published by:*

**DISCOVERY PUBLISHING HOUSE PVT. LTD.**

4383/4B, Ansari Road, Darya Ganj
New Delhi-110 002 (India)
*Phone*: +91-11-23279245; 23253475; 43596065
*Mobile*: +91 9811179893 / +91 9871656464
*E-mail*: discoverybooksindia@gmail.com
orderdphbooks@gmail.com
namitwasan9@gmail.com
*web*: www.discoverypublishinggroup.com

*Printed at:*
Infinity Imaging Systems
Delhi

# Preface

The present title "Encyclopaedia of Cell Biology" has been written for undergraduate and postgraduate students of all Indian universities. It is a text on structural and functional unit of the organism that has been thoroughly checked and ornamented with easy and clear illustrations. Special emphasis has been concentrated on the fundamentals of the subject. All related students will find this well established text eminently suitable for introductory courses. Professional cell biologists, researchers, and students may find it useful for updating, and reference purposes. One of our major concerns of this title is to provide students with a basic understanding of what a cell is and why the cell is the fundamental unit of life. Our approach throughout has been to focus on the major question involved and the experimental approaches utilized in addressing these questions. Each important point is illustrated with an example to help clarify the idea; the numerous line drawings have all been chosen or created to help convey particular points. The result is a text that is easy to read, without resorting to verbal gimmicks or talking down to readers.

To make the work more comprehensive and informative, the author has consulted many authoritative books, research journals, abstracts, monographs etc., so there can be no claim to originality except in the manner of treatment.

The author expresses his thanks to his friends and colleagues whose continue inspirations have initiated him to bring out this book.

The author expresses his gratitude to Mr. Wasan and staff of M/s Discovery Publishing House Pvt. Ltd. for their whole hearted cooperation in the publication of this book.

**Author**

# Contents

# 1

# INTRODUCTION

Much of the discussion in the first fifteen chapters has dealt with an analysis of the molecular basis of cell function and structure. Many of the cellular properties that seemed a few decades ago to be so mysterious as to be beyond comprehension are now clearly understood in straightforward chemical terms. A prime example is the structure, replication, and decoding of genes.

As knowledge of the chemical basis of cell function and structure has increased, so has our curiosity about the evolutionary route by which contemporary cells have come to be what they are. The path of cell evolution has been toward greater and greater efficiency and diversity of cellular operations, and for the most part this has been achieved by the evolution of continually greater complexity in cell function and structure. As a result, even the simplest contemporary cells we know about, the mycoplasmal bacteria, are still chemically very complex organisms.

Some reflections of the evolutionary path are observable in contemporary cells. For example, it is likely that the first eukaryotic cell evolved a long time ago from a progenitor cell that was probably not greatly different in structure from some contemporary prokaryotes. The progenitor cell, called a *progenote*, gave rise to both prokaryotes and eukaryotes.

The progenote has long since disappeared, and the many functional and structural differences that now separate contemporary prokaryotes and eukaryotes attest to a long evolutionary series of intermediate cell types, particularly in the eukaryotic line of descent; little trace of the intermediates has remained. The attempt to understand cell evolution ultimately leads us to the question of the origin of the first cell.

Although we know roughly *when* in the Earth's history the first cell probably arose (more than $3.5 \times 10^9$ years ago), we do not know and probably never will be able to learn in specific terms *how* it arose. Still it is possible that we may gain a general understanding of the origin of the cell. Not many years ago the question of how life originated seemed beyond serious scientific inquiry. The recent great increase in knowledge of the biochemistry and the molecular nature of cell function and structure now permits formulation of specific schemes of how the first cell might have come into existence.

To some extent it is now possible to devise hypotheses about the origin of the cell that can be tested in the laboratory. Particularly important in all of this thinking is the concept of an *organic soup*, a complex mixture of organic molecules that formed and accumulated during the first one billion years after the Earth was formed, and within which the cell subsequently originated. Laboratory experiments designed to test the organic soup concept are discussed later. We will first consider briefly some early ideas about the origin of life.

## SPONTANEOUS GENERATION OF LIFE

The ancient Greek scholars formulated the idea of *spontaneous generation* of life in which insects and other small animals were thought to arise spontaneously from mud or decaying organic matter. This idea persisted throughout the Middle Ages with descriptions of successful demonstrations of spontaneous generation of worms, flies, eels, frogs, and other organisms from mud and decaying materials. For example, mice were believed to arise spontaneously from cheese wrapped in rags and kept in a dark place.

Frogs were said to form from decaying vegetation in ponds. The appearance of maggots in rotten meat seemed a particularly clear case of spontaneous generation. The overthrow of the doctrine of spontaneous generation began in the late 1600s with the experiments of Francesco Redi, an Italian physician. In his most famous experiment he showed that maggots did not appear in rotting meat when the meat was protected from flies.

On meat exposed to the open air, maggots develop from eggs laid by flies. Redi's experiments discredited the theory of the spontaneous generation of animals. However, microorganisms were discovered by Antonie van Leeuwenhoek, the father of microscopy, in 1677, about the same time as Redi's experiments, and the controversy about spontaneous generation shifted from animals to microorganisms. The debate reached a peak in the 1700s and was led by an Englishman, John Needham, and an Italian, Lazzaro Spallanzani. These two men did similar experiments but obtained different results. A solution of organic matter, for example mutton gravy, was boiled in a glass vessel, and the vessel was sealed. Needham sealed his vessels with corks and consistently observed the growth of microorganisms several days after boiling the solution. He concluded that the microorganisms had originated spontaneously out of the organic matter.

Spallanzani boiled the organic solutions longer and sealed them more carefully. In some experiments he sealed vessels containing organic solutions by melting the neck of the glass vessel and *then* boiled the contents. The pressure in the closed vessels was increased by the heating, and therefore the boiling point of the enclosed organic solution was raised. This is equivalent to the modern method of sterilization called *autoclaving*, in which solutions are heated to 120°C using steam to generate a pressure of 15 to 20 lbs per in$^2$.

Spallanzani consistently observed that his organic solutions remained free of microorganisms. He concluded that the microorganisms that appeared in solutions heated and left open came from the air. However, others repeated Spallanzani's experiments and observed the appearance of microorganisms.

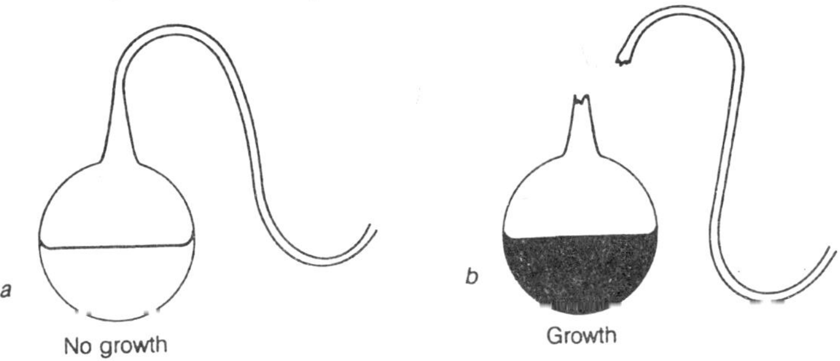


*Figure 1.1: Pasteur experiment about the source of microorganisms that grew in boiled nutrient broth. (a) No growth in broth boiled in a flask with an S-shaped neck. (b) Growth in a flask with a broken neck.*

We must conclude that these experimenters were less careful than Spallanzani and had not achieved sterilisation of their solutions. The outcome was, however, that Spallanzani's experiments did not put an end to the controversy about spontaneous generation of microorganisms. We now know that simple boiling of a solution (Needham's experiment) is insufficient to kill all forms of microorganisms. Many kinds of bacteria and fungi develop into spores when conditions are unfavorable for growth. In the form of a spore the cell is dehydrated, metabolically inert, and enclosed by a thick, tough cell wall. When placed in an environment that is favorable for growth, spores germinate into a metabolically active form and resume growth and reproduction. Spores can survive for many years in a dried state.

They are not killed by brief treatment with boiling water but can be destroyed by prolonged boiling or by autoclaving. Bacterial and fungal spores are everywhere in our environment and are a common component of airborne dust. They are a major nuisance to those who work with animal and plant cell cultures, since a single contaminating spore germinates and grows rapidly in the nutrient media required for cell culture.

A further complication is that autoclaving breaks down some of the components in nutrient media required for growth of animal cells. To avoid the deleterious effects of heat, sterilisation is now often achieved by passing media through an extremely fine filter that removes any microorganisms or their spores. The argument about spontaneous generation was settled to the satisfaction of most scientists only by the experiments of Louis Pasteur in the 1860s.

One of his experiments consisted of sterilising nutrient broth by prolonged boiling in a flask with an S-shaped neck. The flask was open to the air, but airborne spores were trapped in the S-shaped neck, and the broth remained sterile. When the neck of the flask was subsequently broken off, allowing airborne dust to enter, microorganisms began to grow in the broth. Pasteur also showed that microorganisms (presumably in spore form) could be collected by passing air through a filter made of cotton. The disproof of spontaneous generation leaves us with an enigma. If cells can arise only from preexisting cells, where did the first cell on Earth come from?

The experiments by Spallanzani and Pasteur showed that cells could not arise spontaneously from organic matter, at least under the conditions of their experiments. Modern experiments on this problem are designed to test the hypothesis that under certain conditions, particularly those that might have existed on Earth before the appearance of life, a simple primitive cell might have formed spontaneously.

Thus, biologists accept that cells do not now arise spontaneously from organic material, but most believe that given the right conditions and sufficient time (many millions of years), spontaneous generation of a cell did happen at least once, and all contemporary cells have descended from that first cell. In considering the current thinking and experimentation on the origin of life it is necessary to order real and postulated events onto a time scale beginning with the formation of the Earth.

## AGE OF THE EARTH

Scientists have estimated the age of the Earth by measuring the amount of decay of certain radioactive isotopes that has taken place in rocks. For example, radioactive potassium, 40K decays into argon (40A) and calcium ($^{40}$Ca).

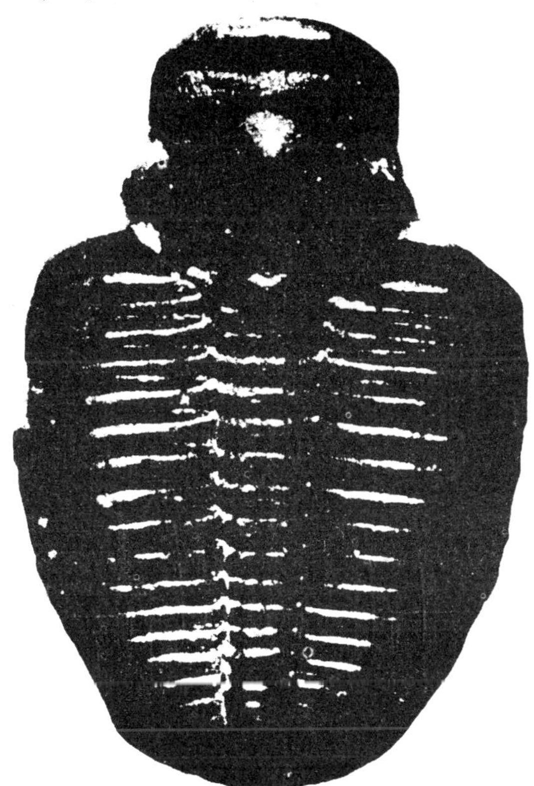

*Figure 1.2 : A fossil trilobite. Trilobites were among the first arthropods, which includes crustaceans, insects, mites, and spiders. Trilobites existed for several hundred million years but died out about 250 million years ago.*

About 90 percent of 40K decays to $^{40}$A and ten percent to $^{40}$Ca. It is safe to assume that argon would not have been present when the rocks formed because this element occurs as a gas and would have been excluded from rocks. Any argon now present in rocks must therefore have derived from the decay of $^{40}$K after their formation. The half-life of 40K is $1.26 \times 10^9$ years. Knowing this, one can compute the age of the particular rock by measuring how much of the $^{40}K$ has de cayed to $^{40}A$, that is, by measuring the ratio of $^{40}A$ to $^{40}K$.

In addition to the potassium-argon method of dating, the age of rocks can be estimated from the ratio of rubidium to strontium or by the uranium-thorium-lead method. The three dating methods all give about the same age for the oldest rocks, $4.6 \times 10^9$ years. This is generally accepted as the age of the Earth. Formation of the Earth is believed to have occurred by aggregation of particles ranging in size from dust to asteroids over a period of millions of years.

The process still continues with the capture from space of dust and meteors by the Earth's gravity. Contrary to earlier views, it is now generally believed that the surface of Earth was not molten when the Earth was formed, which means that the surface temperature remained below 900°C. This is important because it means that simple organic materials, which are known to be present outside the Earth, might have been included during formation of the Earth and survived to become a part of the primitive Earth.

## THE SCALE OF BIOLOGICAL TIME

Until the late 1960s the oldest known fossils had been found in sedimentary rocks about 600 million years old. These represented highly evolved invertebrate animals such as *trilobites*, whose hard shells readily gave rise to fossils. The fossil record indicates that small invertebrates and plants were the only multicellular organisms in existence from about 600 to 500 million years ago.

This interval of 100 million years is called the *Cambrian period.* The Cambrian period was followed by a succes sion of periods that are defined by the fossils of progressively more highly evolved plants and animals, extending to the *Quaternary period,* in which we now live. The period before the appearance of the first fossilised invertebrates, extending from the origin of the Earth 4.6 billion years ago up to the beginning of the Cambrian period 600 million years ago, is called the *Precambrian period.* Six hundred million years is an underestimate of the time of appearance of the first multicellular organisms.

Multicellular organisms that were structurally more simple, lacking hard shells, and leaving no apparent fossil record undoubtedly preceded the appearance of such highly evolved forms as hardshelled invertebrates. Recently, worm tracks have been discovered in rocks at least 100 million years older than fossils from hard-shelled animals, showing that softbodied multicellular organisms were present at least 100 million years before the start of the Cambrian period, or 700 million years B.P.

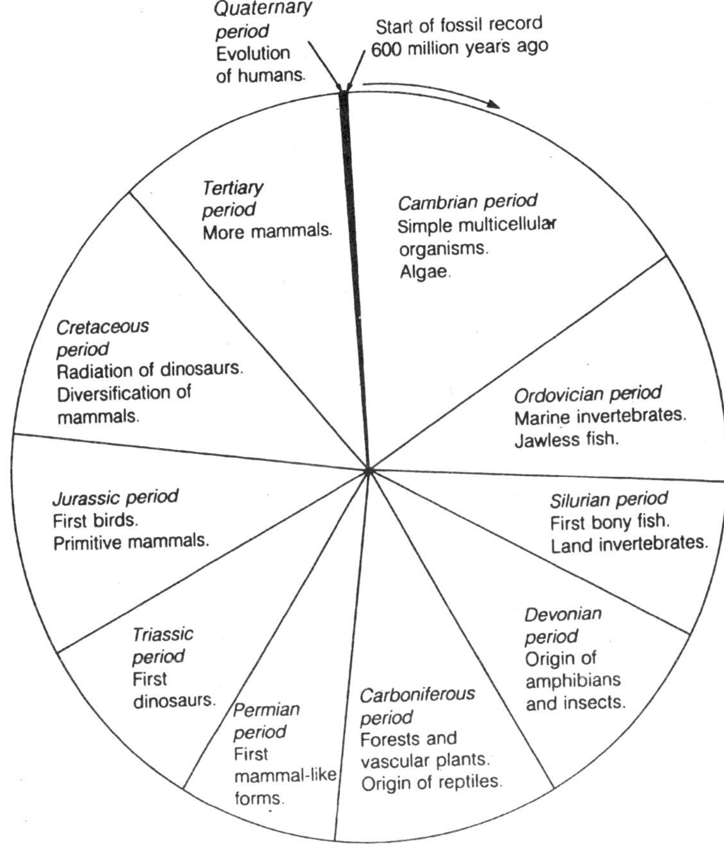


*Figure 1.3 : Major periods and events in the progression of life.*

(before present). We do not know when the first multicellular animals evolved, but it must have been longer than 700 million years ago. Also, however long ago the first multicellular animal evolved, it must have been preceded by an interval in which all organisms were unicellular prokaryotes and eukaryotes. Furthermore, this interval was most likely preceded by a period in which only progenotes existed. All of this was largely supposition until the late 1960s, when the electron microscope was used in a successful search for microfossils of unicellular organisms in ancient rocks.

## MICROFOSSILS OF UNICELLULAR ORGANISMS

Many descriptions of microfossils in rocks formed far back in

Precambrian times have been published in scientific journals in the last 20 years. The oldest microfossils discovered so far are filamentous and spherical structures that strongly resemble bacteria, including blue-green bacteria.

These microfossils occur abundantly in flintlike rocks called chert in South Africa and in rocks of Western Australia, both of which are $3.5 \times 10^9$ years old. Cherts formed in Precambrian times also typically contain 0.5 to 1.0 percent organic material, including substances that could be breakdown products of chlorophyll and proteins. These findings support the interpretation that microstructures such as those in Figure elsewhere in this chapter are indeed fossils of unicellular organisms.

On the basis of these studies it is now generally believed that life existed on Earth at least $3.5 \times 10^9$ years B.P., and that photosynthetic organisms (bluegreen bacteria) had already evolved by that time. For reasons discussed later, it is generally believed that the first cells were highly heterotrophic in their nutrition, obtaining energy and nutrients from an abundance of organic molecules in the environment. Evolution of photosynthetic prokaryotic organisms followed sometime later. Thus, we may conclude that the first cells, which were probably prokaryotic-like, arose sometime before $3.5 \times 10^9$ years ago, less than $10^9$ years after formation of the Earth.

## ORIGIN OF EUKARYOTIC CELLS

The fossil record provides an estimate of when eukaryotes originated. Some of the microfossils found in Bitter Springs chert of Australia contain structures resembling cytoplasmic organelles and nuclei and are believed to represent unicellular eukaryotes. This chert is 900 million years old, and unicellular eukaryotes therefore seem to have been well established by that time.

Fossils of possibly nucleated cells have also been discovered in Beck Spring dolomite in California, which is about $1.3 \times 10^9$ years old. The oldest microfossils that might be remnants of eukaryotic cells have been found in rocks 1.4 to $1.6 \times 10^9$ years old from the Ural Mountains in the U.S.S.R. These microfossils are similar in form to existing kinds of unicellular green algae.

Shale from Montana about $1.4 \times 10^9$ years old has yielded microfossils of filamentous blue-green bacteria as well as much larger spheroidal, thick-walled structures that are believed to be cysts of eukaryotes. From these and similar discoveries it is generally believed that eukaryotes originated at least $1.4 \times 10^9$ years ago. Thus, the origin of preprokaryotes (progenotes) preceded the origin of eukaryotes by about 2.1 × $10^9$ years. Many differences separate contemporary prokaryotes and eukaryotes. They both probably evolved from progenotes, and it is virtually certain that not all the differences arose simultaneously. Modern prokaryotes probably do not differ in structure nearly as much from the progenote ancestor as do modern eukaryotes. The origin of eukaryotes must have occurred with a single change, and this was followed by subsequent evolution of a succession of changes that now make eukaryotes very different from prokaryotes.

We have little idea of which difference represents the first step in divergence of eukaryotes from the progenote. It might have been the acquisition of any one of a number of properties-new kinds of genes, multiple chromosomes, a larger content of DNA, a nuclear envelope, histones, a new principle of gene regulation, and so forth.

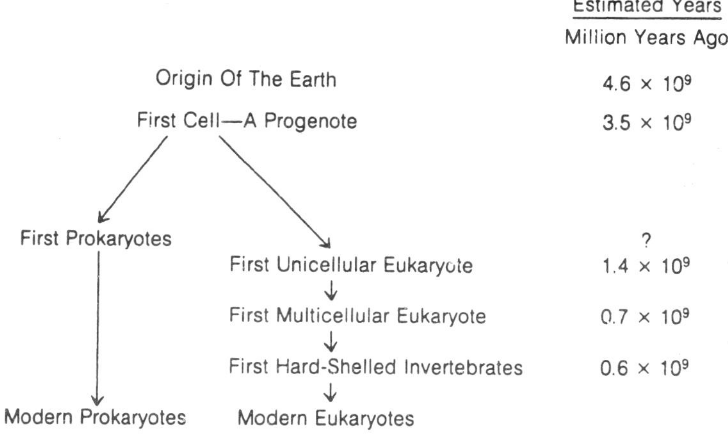


*Figure 1.4: Estimates of timing of appearance of major groups of organisms.*

The cell and molecular biology of only a small fraction of extant eukaryotes has been studied so far, but it is unlikely that any primitive eukaryotic cells have survived to give us insight into the origin of the eukaryotic cell line. Nevertheless, increased knowledge of the function and structure of contemporary eukaryotic cells may someday provide an indication of the origin and evolution of the eukaryotic cell.

Continued study of microfossils will no doubt result in more accurate estimates of the timing of such events as the origin of the first cells (progenotes), the evolution of the prokaryote and eukaryote lines, and evolution of the first simple multicellular organisms from unicellular eukaryotes. For the present the best estimate of the time scale of events may be summarised as shown in Figure elsewhere in this chapter. Once the first cell had formed $3.5 \times 10^9$ or more years ago, we can understand, at least in principle, how the subsequent course of events was dictated by the evolution of an ever-increasing complexity in cell functions and structures. We know little about the particular steps, such as the evolution of regulatory genes, the evolution of photosynthesis, the evolution of the first eukaryotic cell, or the evolution of those genetic mechanisms that made possible the first multicellular organisms.

Future research in molecular biology, genetics, and cell biology may yet give us a better idea of these processes of cellular evolution. However, we are faced with a conceptually far more difficult problem than cellular evolution, and that is the matter of how the cell came into existence in the first place. It is impossible to formulate any reasonable scheme by which a cell might have formed *directly* from the *inorganic* materials present on the primitive Earth.

The jump from inorganic chemicals to organic molecules capable of self-replication is simply too enormous. A solution to this conceptual dilemma was first proposed by the Russian biochemist, Alexander I. Oparin, and the British biologist J.B.S. Haldane begin' ning in the 1920s, and is now generally called the *organic soup concept.*

## THE ORGANIC SOUP THEORY

The great contribution of Oparin and Haldane to the subject of the origin of life was based on the idea that in the period before life arose the atmosphere of Earth contained hydrogen ($H_2$), methane ($CH_4$), ammonia ($NH_3$), and water ($H_2O$), but no free oxygen ($O_2$). Thus, Oparin and Haldane proposed that the prelife atmosphere of Earth was highly reducing in a chemical sense.

From a variety of evidence, geologists, cosmologists, and chemists now generally agree that primitive atmosphere was chemically reducing in nature. As an example of these lines of evidence, early Precambrian rocks contain ferrous iron, which is unstable in the presence of $O_2$. Therefore, the early Precambrian rocks must have been laid down in the absence of atmospheric $O_2$.

Oparin and Haldane both reasoned that a reducing atmosphere consisting of $H_2$, $CH_4$, $NH_3$, and $H_2O$ would be favorable for the spontaneous formation of simple organic molecules, and these might then polymerise spontaneously into macromolecules. These macromolecules might then accumulate in the oceans and lakes of the time, giving rise to *organic soups*. It is doubtful that the accumulation would have been great because many kinds of organic molecules are unstable in aqueous solutions and are slowly and spontaneously hydrolysed.

However, one may suppose that organic molecules may have become concentrated by adsorption to solid surfaces (a common phenomenon) or through the rapid evaporation of lakes.

It was Oparin's idea that the first cell arose not from inorganic substances but from a mass of prebiologically formed organic material. Oparin's idea remained unknown in the West for a long time, probably because it was published in Russian. In 1938 Oparin's book entitled *The Origin of Life* was published in English, and immediately his theory attracted wide attention.

Fifteen years later, in 1953, the Oparin-Haldane proposal about the spontaneous formation of organic molecules was tested directly. With the apparatus shown in Figure elsewhere in this chapter, a gaseous mixture of $H_2$, $CH_4$, $NH_3$, and $H_2O$ was exposed to electric spark discharges. Water was first added to the flask. The air was pumped out with a vacuum pump and the apparatus was then filled with a mixture of hydrogen, methane, and ammonia.

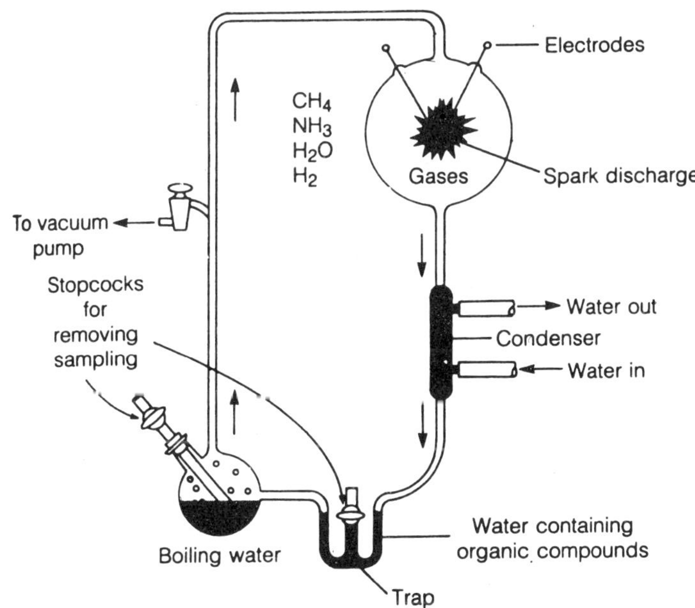


*Figure 1.5: Appaıatus for synthesis of amino acids and other organic compounds by spark discharges.*

The electric discharges, which were common in the primitive atmosphere (as lightning), provided energy for the synthesis of molecules from the four starting components. The water in the flask was boiled to cause circulation through the apparatus and remove any reaction products from the spark zone. Reaction products collected in the condensing water in the condenser and accumulated in the water phase. The experiment was run for one week, and the water then analysed for any organic compounds that might have formed.

The results of the analysis, which are shown in Table elsewhere in this chapter, are astonishing in at least two respects. A complicated mixture of small amounts of hundreds or even thousands of compounds is theoretically possible and might have reasonably been expected. Instead a small number of compounds accounted for *most* of the reaction products.

Second, the molecules produced included several of major biological importance, particularly the amino acids glycine, alanine, aspartic acid, and glutamic acid. Fifteen percent of the carbon added (as $CH_4$) to the apparatus was recovered in the compounds identified in Table elsehwhere in this chapter. Additional carbon was converted into unidentified, tarlike, high-molecular-weight, organic polymers.

This experiment has been repeated with various modifications many times in other laboratories. As a result of these experiments almost all 20 amino acids, as well as purines, pyrimidines, ribose, nucleosides, and nucleotides have been produced abiologically under simulated early-Earth conditions.

In recent years radioastronomy has provided evidence by microwave spectroscopy that abiological synthesis of large quantities of biologically important molecules occurs commonly in the universe outside the Earth. These molecules include $H_2$, $H_2O$, $NH_3$, $H_2S$, CO, and HCN and the organic molecules cyanoacetylene ($C_2HN$), methanol ($CH_3OH$), ethanol ($CH_3CH_2OH$), formaldehyde ($CH_2O$), formic acid (HCOOH), formamide ($HCONH_2$), acetonitrile ($CH_3CN$), and acetaldehyde ($CH_3CHO$).

This list is striking because it includes the very compounds that are the most important for the abiological synthesis of amino acids, purines, pyrimidines, and sugars. For example, formaldehyde, acetaldehyde, and hydrocyanic acid (HCN) react to form glutamic acid. Cyanoacetylene is a precursor of pyrimidines, and in particular can form a large amount of cytosine. Aldehydes, HCN, and $NH_3$ yield a variety of amino acids. Ribose, glucose, and other sugars are formed spontaneously in an alkaline solution of formaldehyde. Hydrocyanic acid is a precursor of glycine and purines.

A group of meteorites known as *carbonaceous chondrites* also contain organic molecules. The *Murchison meteorite*, which fell near Murchison,

Australia in 1969, contains two percent carbon, much of which is present as a complex mixture of organic molecules. Among these are glycine, alanine, valine, proline, glutamic acid, and aspartic acid. Similar findings have been made with the *Murray meteorite*, which fell in the U.S. in 1950. Pyrimidines and possibly purines are also present in the Murchison meteorite.

**Table 1.1 : Some Organic Molecules Formed from $CH_4$, $NH_3$, $H_2O$, and $H_2$ as a Result of Spark Discharges**

| | |
|---|---|
| Glycine | Succinic acid |
| Glycolic acid | Aspartic acid |
| Sarcosine | Glutamic acid |
| Alaninc | Iminodiacetic acid |
| Lactic acid | Iminoaceticpropionic acid |
| N-Methylalanine | Formic acid |
| α-Amino-n-butyric acid | Acetic acid |
| α-Aminoisobutyric acid | Propionic acid |
| α-Hydroxybutyric acid | Urea |
| β-Alanine | N-Methyl urea |

The carbonaceous chondrites condensed from the same gaseous mass (the solar nebula) from which the sun and planets condensed. It is likely, therefore, that the organic molecules found in the meteorites were formed in the gaseous nebula.

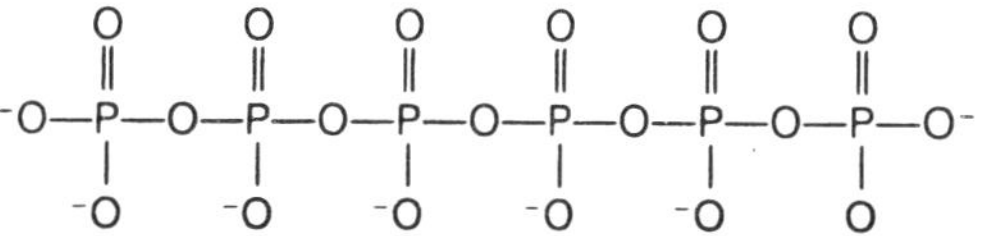


**Polyphosphate**

*Figure 1.6 : Polyphosphate molecules contain variable numbers of phosphate groups. Hydrolysis of polyphosphate molecules can be coupled to peptide bond formation between amino acids, a dehydration-condensation reaction.*

The Earth was also formed from dust and asteroids that condensed from the same gaseous mass, suggesting that organic molecules may have been present from the very beginning of.the Earth's history. Of course, the presence of or ganic molecules in the solar nebula raises the possibility that organic molecules are present in other gaseous masses throughout the universe.

In sum, the synthesis of organic molecules under simulated conditions of prelife Earth, the detection in other parts of the universe of substances that spontaneously react to form organic molecules, and the presence of organic molecules in meteorites all add plausibility to Oparin's idea that the origin of the first cell was preceded by the formation and accumulation of large amounts of organic molecules of major biological importance.

## PREBIOLOGICAL FORMATION OF MACROMOLECULES

With prebiological synthesis of such monomers as amino acids, purines, pyrimidines, nucleotides, and sugars the origin of life no longer seems so inscrutable. However, to go from monomers to a cell, no matter how simple and primitive, is still a big leap. We now know, however, that under the appropriate conditions, a solution of amino acids can polymerise into large polypeptides, and nucleotides can polymerise into nucleic acid molecules.

Thus, it is generally supposed that accumulation of monomers was followed by the prebiological formation of macromolecules and that the first cell arose by an aggregation of macromolecules. What is required for abiological formation of macromolecules? Remember that the polymerisation of amino acids into polypeptides and of nucleotides into polynucleotides occurs by dehydration-condensation of the monomers.

To form a peptide bond between two amino acids (condensation) a molecule of water must be removed (dehydration). Similarly, water is removed in the formation of the 3',5'-phosphodiester bond between two nucleotides. These polymerisations do not occur readily in aqueous solutions because the presence of many water molecules opposes dehydration, driving the reaction toward depolymerisation (hydration of the monomers) rather than polymerisation.

The problem has been solved in the cell by coupling the hydrolysis (bond breakage by addition of water) of ATP to the dehydration-condensation of amino acids or nucleotides into polymers. The dehydration is accomplished by forming an aminoacyl-tRNA at the expense of ATP in the case of peptide bond formation, and by splitting $PP_i$ from nucleoside triphosphate in the case of nucleic acid synthesis.

The overall process, however, is a coupling of water removal from monomers to water addition (hydration) to ATP (to form ADP + $P_i$ + $H^+$). It is highly unlikely that ATP was present in a sufficient amount in the prebiological environment to drive polymerisation reactions. However, *polyphosphates* (the simplest polyphosphate is pyrophosphate) can bring about polymerisation of amino acids by a dehydration reaction in the same manner as does ATP, and polyphosphates could readily have

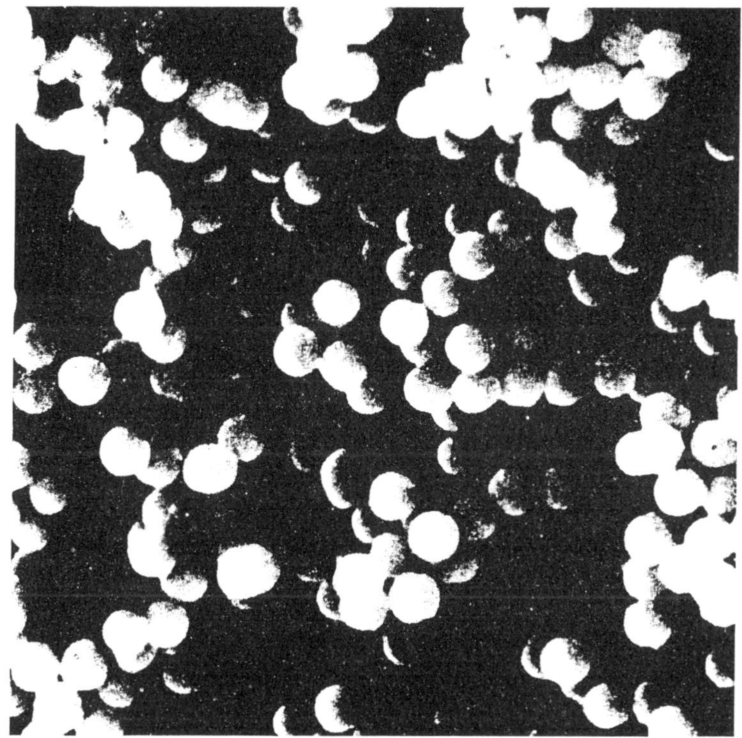

*Figure 1.7: Scanning electron micrograph of proteinoid microshperes. Each is about 2 μm in diameter.*

formed under prelife conditions. Polyphosphates can form when orthophosphate ($PO_4^{3-}$) is warmed in the presence of urea and $NH_4^+$ (urea is a product of the organic soup experiment).

Polyphosphates have been used experimentally to promote the synthesis of AMP from adenine, ribose, and phosphate, and to drive the formation of polynucleotides. Other compounds might also have served as dehydration-condensation agents.

These compounds include carbodiimide, cyanate, cyanogen, and cyanovinyl phosphate, all of which might readily have formed under prelife conditions. Monomers can also be caused to polymerise by adsorption on the surface of certain minerals such as clay and apatite compounds. Adsorption of amino acids to one common clay known as montmorillonite is followed by condensation of the amino acids into long polypeptide chains.

Another way to bring about polypeptide formation is to heat a mixture of dry amino acids to 130 to 180°C for a few hours. In the absence of water the necessary dehydration reaction is strongly favored. Large complex polypeptides formed by this method aggregate into microspheres when mixed with water. Protein-like molecules form when aminoacyl adenylates are mixed.

These aminoacyl adenylates are the activated form of amino acids used in cellular protein synthesis. Aminoacyl adenylates readily form abiologically when AMP and amino acids are mixed in the presence of dehydration-condensation agents. Not only do the abiologically formed protein-like ,polymers form stable microspheres but, remarkably, such polymers possess low levels of catalytic power, for example, in the decarboxylation of pyruvate to acetaldehyde and $CO_2$.

Thus, at least some of the protein-like polymers qualify as primitive enzymes.

## FROM MACROMOLECULES TO CELLS

The strong probability that polypeptides and polynucleotides formed under prelife conditions further reduces the difficulties in conceptualising how life originated. Yet we are still faced with formidable problems in trying to understand the origin of the cell.

Abiologically formed macromolecules may aggregate into cell-like structures, such as the proteinoid microspheres in Figure 16-8, and these may grow and subdivide. What additional properties must an aggregate of macromolecules possess to be called a cell? The answer to this is not so easy as it might seem initially.

It seems essential that a genetic mechanism be present that allows the primitive cell to duplicate itself precisely. It is difficult to imagine how this could be achieved without a template-copying mechanism of the type present in nucleic acid replication. We may also add the requirement that the macromolecular components must have functions, however crude, that are useful for survival and reproduction. Enhancement by enzymatic catalysis of polymerisation of monomers into more macromolecules would be one such funcinon.

These functions must be inherited through cell reproduction. The presence of a genetic inheritance carries with it the potential for chance improvement of macromolecular functions through mutation. Once genetically inherited functions have been acquired, the way is open for cellular evolution.

We might suppose that the first cell had to contain nucleic acids because only nucleic acids are known to replicate themselves with a reasonable amount of precision and therefore constitute a mechanism of genetic inheritance. As already stated, the formation of nucleotides under simulated prelife conditions has already been accomplished in the laboratory. In addition, these nucleotides can be caused to polymerise into short polynucleotides when a dehydration-condensation agent is added.

With an appropriate polymerase obtained from cells, a single-stranded DNA or RNA chain acts as a template for the synthesis of a complementary chain from nucleoside triphosphates. Essentially the same can be accomplished without a polymerising enzyme, that is, under prelife conditions. For example, oligomeric chains of thymidylate residues can be generated under prelife conditions. In turn, oligomers composed of six thymidylate residues (hexamers) have been found to join together slowly to form longer polynucleotides of thymidylate residues in the presence of a dehydration-condensation agent.

If polyadenylate is added to the mixture, the rate of joining of the thymidylic acid hexamers is greatly speeded up; apparently the polyadenylic acid acts as a template that binds the hexamers of thymidylate residues by base-pairing and thereby aligns the hexamers so that end-to-end joining takes place much more rapidly. In bringing together hexamers of thymidylate residues, and by promoting their end-to-end ligation, polyadenylate is actually functioning as a catalyst. Similarly, hexamers of polyadenylate residues may be coupled together into longer polynucleotides on a template of polyuridylate residues. Polycytidylate chains act as templates for polymerisation of GTP into chains up to 40 nucleotides long.

These experiments give evidence that polynucleotides might well be able to replicate themselves in the familiar template fashion without catalysis by protein enzymes. Therefore, it seems possible that cellular reproduction was from the beginning based on polynucleotide replication. We know of no way in which proteins can function as templates to guide their own reproduction.

Not long ago, nucleic acids were thought to lack the kind of catalytic capabilities necessary to give the first cell its first functions. Thus it seemed that proteins (primitive enzymes) must have been present early in primitive cells. The supposition that proteins had to be present to provide enzymatic activity in primitive cells has changed dramatically as a result of recent studies on excision of introns from primary RNA transcripts in eukaryotes. The precursor ribosomal RNA transcript in *Tetrahymena* undergoes catalytic self-splicing in which the RNA intron acts as an enzyme to remove itself from the transcript by cutting and

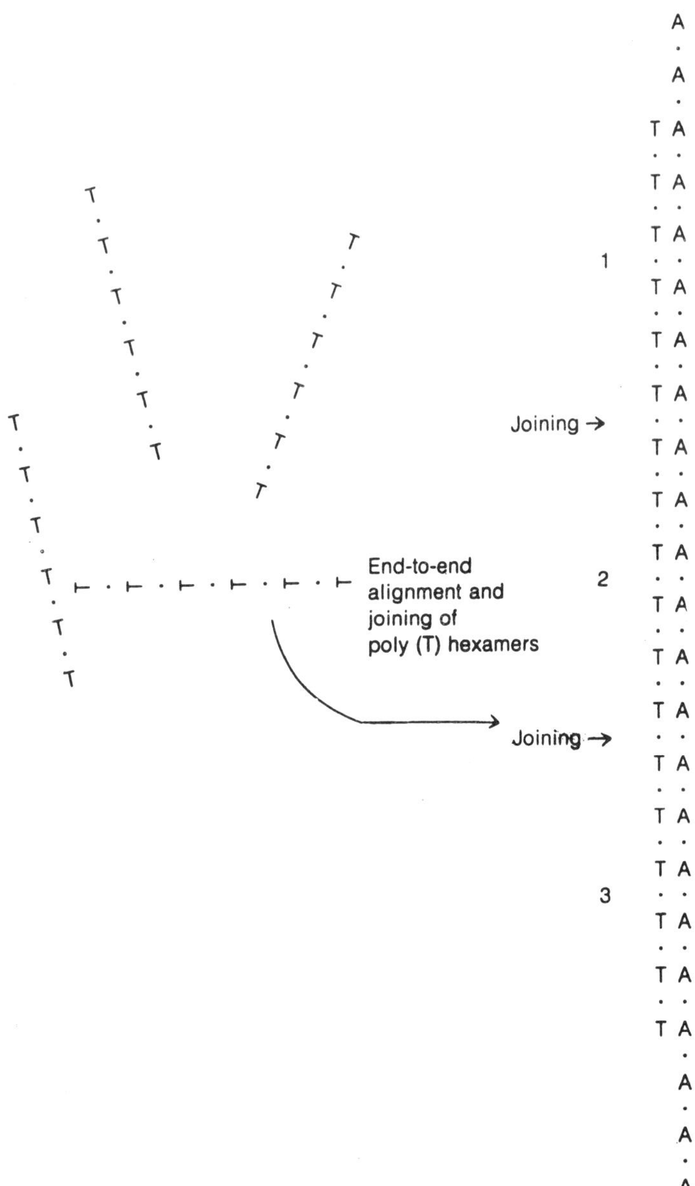


*Figure 1.8: Hexamers of poly(T) aligned end-to-end on poly(A) and joined into long molecules of poly(T).*

splicing together of RNA ends.

Moreover, the intron RNA can act as a ribonuclease to remove nucleotides from RNA, and astonishingly the intron RNA can also work as a polymerase, catalysing the formation of polycytidylate chains on a poly(G) template that it itself contains. Segments of RNA such as the intron of precursor rRNA in *Tetrahymena* have been given the name *ribozymes*. These discoveries favor strongly the idea that the first cells used RNA molecules both to hold genetic information and as enzymes to catalyse RNA metabolism.

The subsequent switch from RNA to a DNA double helix as the genetic material is not a conceptually difficult problem. However, how self-catalysed RNA replication evolved into the current genetic mechanisms in which information in nucleotide sequences is translated into amino acid sequences via tRNA adaptors is much harder to envisage.

## THE FIRST CELL: AN EXTREME HETEROTROPH

We cannot reasonably suppose that the first cell was capable of more than a few enzymatic activities. Hence, it must have depended on abiologically formed organic molecules in its environment for growth and reproduction. With the evolution of anabolic capabilities, stringent dependence on heterotrophic existence presumably decreased; indeed, depletion of essential components (e.g., amino acids and nucleotides) in the environment necessitated the evolution of anabolic activities.

Eventually, the exogenous supply of nutrients such as nucleotides and amino acids would begin to be depleted. Any primitive cell that acquired through chance the ability to catalyse the formation of one or another nucleotide or amino acid from some closely related molecule still available in the environment would have an advantage over its contemporaries. Hence early cellular evolution surely moved in the direction of *autotrophy*. It seems unnecessary to postulate that in its earliest form the cell possessed an enveloping membrane.

However, as soon as anabolic and energy-generating capabilities began to be acquired, a membrane became important for retention of valuable metabolic products. Hence, we may suppose that a plasma membrane of some sort appeared early in the evolution of metabolism.

## EVOLUTION OF ENERGY-GENERATING MECHANISMS

The acquisition of anabolic capabilities carries with it a requirement for energy to drive synthesis of molecules. The primitive environment may have contained some energy-rich substances (e.g., polyphosphates), but these would probably have been quickly used up. Therefore, the ability of the cell to generate its own energy supply probably occurred early in cell evolution.

For several reasons, it seems likely that the first energy generating mechanism was a primitive form of glycolysis. All contemporary cells possess glycolysis, which indicates that it evolved before prokaryotic and eukaryotic cells diverged from a common ancestor. Glycolysis is also mechanistically simpler than photosynthesis and presumably preceded photosynthesis in evolution. Since the early atmosphere of Earth lacked oxygen, respiration was impossible.

In fact, it is generally believed that the oxygen now present in the atmosphere is the result of photosynthesis. Hence, respiration probably evolved after photosynthesis. The order of evolution of energy-generating mechanisms is therefore believed to have proceeded from glycolysis to photosynthesis to respiration. We can summarise the discussion about the origin and early evolution of the cell by elaborating on the scheme presented earlier.

## ORIGIN OF THE CHLOROPLAST AND THE MITOCHONDRION

The origin of these organelles represents one aspect of eukaryotic cell evolution about which we have gained some understanding. The mitochondrion is believed to have arisen through a symbiotic association between an early eukaryotic cell and a bacterium. According to this hypothesis a bacterium was engulfed by the cytoplasm of a eukaryotic cell, where it became a permanent resident, growing and reproducing with little if any detriment to its host.

Such arrangements are not uncommon among contemporary fungi and protozoa. *Amoeba proteus,* for example, harbors a bacterium that grows and multiplies in the cytoplasm of the amoeba. A single amoeba contains thousands of these bacteria. The bacterium has become adapted to the environment provided by the amoeba cytoplasm to the extent that it can no longer grow and divide outside the amoeba.

Whether the bacterium contributes anything to the well being of the amoeba (a symbiotic relationship) or whether it is strictly a parasite is not known. The arrangement now present in *A. proteus* may resemble an early stage in mitochondrial evolution. The relationship between the early eukaryotic cell and its acquired bacterium evolved into a symbiotic one in which the bacterium became specialised for the respiratory synthesis of ATP, which it supplied to its host. The eukaryotic host, in turn, evolved in the direction of providing more and more of the proteins necessary for the structure and function of the symbiont, and this was accompanied by a corresponding loss of genes (DNA) in the symbiont. Assuming that the host cell was originally aerobic, it lost its own enzymatic machinery for respiration since all respiration is carried out in mitochondria in contemporary cells.

The genetic and biochemical interrelationships between mitochondria and the rest of the cell are intricate and complex. Chloroplasts are also likely to have evolved from blue-green bacteria acquired by an early eukaryotic cell. According to the hypothesis about the symbiotic origin of the mitochondrion and the chloroplast, the DNA in these organelles represents all that remains of the original chromosome of an aerobic bacterium or blue-green bacterium. The strength of the symbiosis hypothesis lies in basic similarities between the organelles and free-living bacteria.

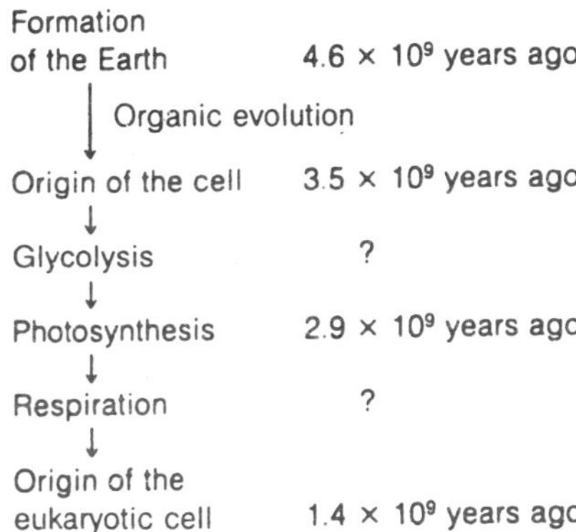


*Figure 1.9 : Estimate of occurrence of major cellular phenomena.*

First, the DNA of mitochondria and chloroplasts does not have histones associated with it, and neither does the DNA of bacteria or blue-green bacteria. The ribosomes of mitochondria and chloroplasts correspond in size to prokaryotic ribosomes rather than eukaryotic ribosomes.

The rRNAs of organellar ribosomes, which are coded for by organellar DNA, are about the same size as prokaryotic rRNAs and hence are smaller than eukaryotic rRNAs. Moreover, organellar rRNAs have nucleotide sequences that are present in prokaryotic rRNAs but not in eukaryotic rRNAs.

Finally, the drug chloramphenicol inhibits protein synthesis in bacteria and in mitochondria, but has no effect on protein synthesis in the cytoplasm of eukaryotic cells. Thus, mitochondrial and bacterial protein synthesis share a basic property that is not present in eukaryotic protein synthesis. There is a variety of other, corroborating observations involving similarities in other organellar and bacterial properties, but the above are the main ones.

# 2

# CELL ORGANISATION

Although the term cell was first used in biological literature by Robert Hooke in 1665 to describe a compartment seen in the structure of cork, the cell as a unit of structure and function of all living things was not recognised until about 1839 when Schleiden and Schwann developed the *cell doctrine.* In modern terms the cell may be defined as "a unit of biological activity delimited by a selectively permeable membrane and capable of self reproduction in a medium free of other living systems".

It is common knowledge that the cell is differentiated into *nucleus* and *cytoplasm,* each of which is surrounded by a membrane, the nucleus by the *nuclear envelope* and the cytoplasm by the *cell* or *plasma membrane.* The outer portion of the cytoplasm, called the *ectoplasm, is* generally more finely granular and of a stiffer consistency than the inner portion, called the *endoplasm.* It is familiar to all who have studied plant and animal cells that they are diverse in size and form.

A brief summary of the range in size of cells is given in Table elsewhere in this chapter and of the range in form in Table elsewher in this chapter. It is of moment, perhaps, to point out that cells are examples of a remarkable degree of miniaturisation. Thus the human sperm carries in its minute mass half of all the genetic determinants of a mature individual.

The large cells listed in Table elsewhere in this chapter, which might appear to be exceptions to miniaturisation, contain a vast store of nutrient for development of the embryo. The *Valonia* cells have a large central sap vacuole as well as a massive cell wall. In both cases the nucleus and cytoplasm represent a very small part of the total cell mass. Although diverse in size and form, cells have a similar intracellular organisation and usually possess the same formed structures or organelles.

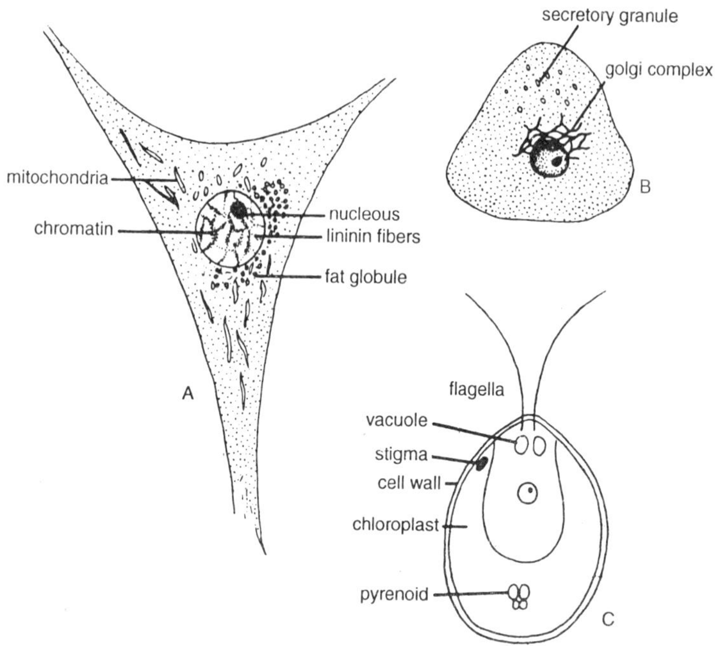


*Figure 2.1: Cells and cell organelles. A is a fibroblast, B is a cell of the pancreas and C is the flagellate Chlamydornonas, at the center of which is a nucleus.*

The presence of most of the organelles in all cells suggests that each organelle performs some fundamental role in the economy of the cell. Determination of the functions of these organelles is one of the ultimate aims of the cell physiologist.

**Table 2.1:Range in Size of Cells.**

| Mass in Grams | Organisms or Cell |
|---|---|
| $10^3$ to $10^2$ | Dinosaur eggs, ostrich eggs, cycad ovules |
| $10^1$ | *Valonia macrophysa* (mature) |
| $10^{10}$ | *Valonia ventricosa* (mature) |
| $10^{-1}$ | *Nitella* (large internode) |
| $10^{-2}$ to $10^{-3}$ | Frog eggs |
| $10^{-4}$ | Striated muscle of man |
| $10^{-5}$ | Human ovum |
| $10^{-6}$ | Large *Paramecium* |

| | |
|---|---|
| | Large sensory neuron of dog |
| $10^{-7}$ | Average *Vorticella* |
| | Human smooth muscle fiber |
| | Human liver cell |
| $10^{-8}$ | Dysentery ameba |
| $10^{-9}$ | Frog erythrocyte |
| | Malarial parasite |
| | Human sperm |
| | Smallest Protozoa *(Monas)* |
| $10^{-11}$ | Anthrax bacillus |
| $10^{-12}$ | Tubercle and pus bacteria |
| $10^{-14}$ | Smallest bacteria |
| | (Limit of microscopic vision) |
| $10^{-16}$ | Filter passing viruses, etc. |
| $10^{-17}$ | *Mycoplasma laidlawii* |

## ORGANELLS SEEN IN FIXED AND STAINED CELLS

Under the microscope, a nucleus is seen in every stained* cell although in bacteria and blue-green algae it lacks a membrane and is less well organised than in true cells of higher forms of plant and animal life. Within the nucleus a *nucleolus* is commonly present and so is a *linin* network bearing the *chromatin*. Between these materials is the *nucleoplasm*. The entire nucleus is surrounded by a nuclear envelope. Just outside the nucleus of animal cells is a *centriole*, a small granule which separates into two before mitosis (cell division). At mitosis the two centrioles move apart and form the poles of the spindles and the centers of the asters.

Scattered in the cytoplasm are *mitochondria*, or chondriosomes, which have been seen in many kinds of cells examined. The appearance of mitochondria depends upon the handling of the tissues before preparation; they may appear as long rods, short rods or small granules.

The *Golgi complex,* named after its discoverer, is particularly prominent in nerve and secretory cells of animals and seems to be closely associated with secretory activities. In fixed and stained preparations it appears to be a canalicular system closely associated with the mitochondria. It is absent from bacteria and blue-green algae. Its true structure and function have been the subject of long controversy which is not yet resolved.

Table 2.2. Types of Cells and Their Uses in the Study of Cell Physiology

| NAME | TYPICAL SHAPE | EXAMPLES | USE IN THE STUDY OF CELL PHYSIOLOGY |
|---|---|---|---|
| | | ANIMAL | |
| Epithelial | Cuboidal or brick-like | Epidermis, glandular lining | Permeability, properties of protoplasm |
| Muscular | Spindle | Smooth, striated | Contractility, metabolism |
| Nervous fibers | Cell body with long motor neuron, | Sensory neuron, structure of protoplasm | Irritability, |
| Connective | Spheroid cells in a mass of extra-cellular material | Cartilage cells, cells in bone and connective tissue | Production of extra cellular secretions |
| Blood | Disk-shaped or ameboid cells in fluid | Red blood cell, leucocyte | Permeability, ameboid move -ment, response to chemicals |
| Eggs | Usually spherical | Sea urchin egg, frog egg | Structure of proto plasm, properties of protoplasm, permeability, function of nucleus, cell division |
| Sperm | Usually flagellated cells | Sea urchin sperm | Function of nucleus, nucleoprotein |

| | | | |
|---|---|---|---|
| Protozoans | Diverse | *Amoeba, Paramecium* | Ameboid and ciliary movement, nuclear function, permeability, effects of environment |
| | | **PLANT** | |
| Epidermal | Cuboidal or brick-like | Epidermis | Permeability |
| Vascular | Elongate (some strengthened by spirals, discs, etc.) | Phloem (in xylem only cell walls remain) | Structure of proto plasm (phloem) |
| Supporting | Elongate, heavy walls | Bast fibers | |
| Undifferentiated (parenchyma) | Spheroidal | Mesophyll, interstitial cells of stem, root, etc. | Photosynthesis, structure of protoplasm |
| Unicellular algae | Diverse | *Chlorella* | Photosynthesis, growth and cell division, circadian rhythms |
| | | **PLANT** | |
| Unicellular fungi | Diverse | Yeast | Metabolism, heredity |
| Filamentous fungi | Elongate cells | *Neurospora* | Biosynthesis, heredity |
| Bacteria | Diverse | The colon bacillus | Biosynthesis, metabolism heredity |

*Lysosomes* are membrane-bound organelles containing hydrolytic enzymes. It is likely that the cell interior is protected from the enzymes by such packaging. The packaging may result from the formation of the enzymes on the surface of the endoplasmic reticulum and accumulation within it. *Plastids* are cytoplasmic organelles which may or may not contain pigment (for example, chlorophyll is the green pigment in chloroplasts, one of the kinds of plastids in green plants).

*Vacuoles,* such as the contractile vacuoles of protozoans, and the large central sap vacuoles which are characteristic of plant cells, are often seen in cells. When a cell is cleared by centrifugation of visible granules, oil globules and various organelles, a clear matrix material remains which is thought to represent the fundamental architecture of the cell.

In fixed and stained preparations under the ordinary microscope this substance is devoid of structure, but under the electron microscope, which resolves much finer detail, the matrix material appears to contain a network called the *endoplasmic reticulum.* Upon this reticulum can be seen fine granules called the *ribosomes.* Intercellular cytoplasmic connections called *plasmodesmata* may be found between cells in some tissues. It is of interest to note that although a cell contains but one nucleus it may have about 500 mitochondria, $5 \times 10^5$ ribosomes and probably $5 \times 10^8$, enzyme molecules.

A green plant cell will probably have about 50 chloroplasts. This multiplicity of units is a characteristic feature of cell organisation. Even the nucleus of a diploid organism has two chromosomes of a kind and two genes of a kind, one in each of the members of a homologous pair of chromosomes. As most of the organelles just listed were first described in fixed and stained preparations, the possibility was suggested that the fixing process, by precipitating the proteins, produces artificial structures, or *artifacts,* which do not represent the true structures of the cell.

At the turn of the century Hardy and Fisher showed that some of these organelles could be duplicated in gelatin by adding to it such fixing agents as tannic acid, chromic acid or the salts of heavy metals. Therefore, it became necessary to determine whether the structures observed in fixed cell preparations were actually present in living cells. Many new and ingenious techniques were devised to demonstrate them, and the evidence for each organelle follows.

## EXPERIMENTAL EVIDENCE FOR CELL ORGANELLES IN LIVING CELLS

When a living cell is examined under an ordinary light microscope, its main or ganisation (nucleus, cytoplasm, plastids, granules, edges of membranes) is visible in outline because of differences in refraction of light by these structures.

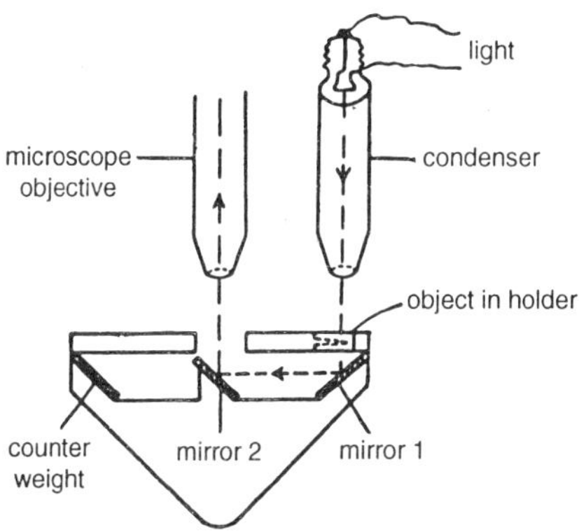


*Figure 2.2: Optical system for ultracentrifuge microscope. The condenser throws an image of the incandescent light on the object.*

When the living cell is centrifuged and observed during the process through the centrifuge microscope, the organelles, differing in specific gravity and size, are seen to move as discrete units.

For example, in centrifuged eggs of the sea urchin *Arbacia,* the red pigment granules are thrown to the bottom of the cell; above them the yolk is deposited and then a layer of mitochondria, and finally the oil droplets are collected at the top of the egg, forming an oil cap. Between the zone of the mitochondria and the oil cap is a clear zone of fine granular protoplasm called the ground substance, within which the nucleus of the egg comes to lie.

Clearly, the specific gravity of the nucleus is less than that of some of the other cellular constituents. The ground substance represents the cytoplasm cleared by centrifugation of the structures suspended in it. No structure in the ground substance of a living cell is visible under the ordinary light microscope, the ultraviolet microscope or the phase microscope.

However, if cells are first fragmented and then centrifuged, the fragments of higher specific gravity settle more rapidly than those of lower specific gravity, other factors being equal, and larger particles settle more rapidly than smaller ones of the same specific gravity. The various particulates within the cell can then be segregated from one another. Large structures such as nuclei are readily separated from other constituents, as are also mitochondria and large granules. After

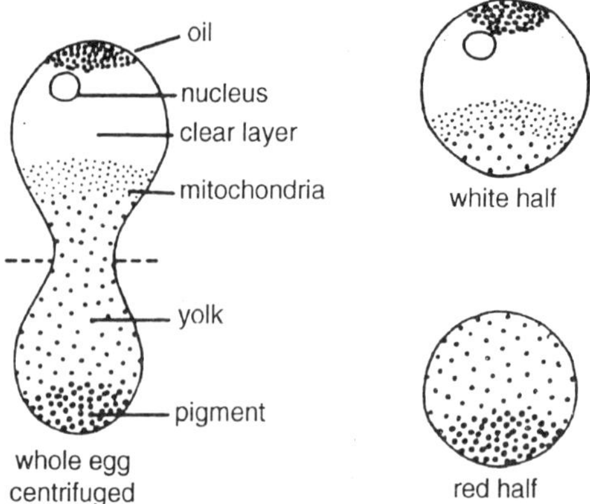


*Figure 2.3: Stratification in a centrifuged egg of the sea urchin Arbacia punctulata.*

all these structures have been removed, minute submicroscopic bodies remain which can be centrifuged down at very high speeds. These are called the *microsomes,* and some of them are ribosomes, the granules seen under the electron microscope in the endoplasmic reticulum. Others are fragments of the endoplasmic reticulum.

**The Nucleus and its Constituents**

The internal organisation of the resting nucleus in a living cell is difficult to discern by ordinary microscopy, and even during mitosis and meiosis the chromosomes are seen only under rare circumstances. However, under the phase microscope, where slight differences in refractive index between diverse cellular constituents are exaggerated, the chromosomes of some cells may be seen quite clearly. In fact, time-lapse photography permits continual observation of a single chromosome preparation throughout a cycle of meiosis or mitosis, corroborating what was previously deduced from a random sampling of many fixed and stained preparations. Furthermore, chromosomes are seen without the distortion which usually attends fixation. Time-lapse photography makes evident not only the sequence of events, but also the dynamic nature of the process-the movements of cell organelles such as mitochondria, and the changes in shape of the entire cells.

Chromosomes are also discernible in ultraviolet photomicrographs of cells in mitosis or meiosis, because the chromosomes absorb short wavelengths of this radiation more strongly than does the cytoplasm. Much has been learned by the use of this technique but it is not as generally useful as phase microscopy because cells are damaged by such radiation, and prolonged examination is thus impossible.

Chromosomes when visible may be reached by micromanipulator needles and moved about in the nucleus or in the cell, indicating that they are discrete units. Moreover, fragmentation and irregularities in the partition of chromosomes can be observed in cells irradiated with ionising radiation. Chromosomes have also been released from a suspension of nuclei (obtained by fractional centrifugation) and they maintain their identity through a variety of manipulations. At present no doubt exists that chromosomes are real structural units, even though they are difficult to identify in the resting stages of cells.

***Mitochondria***

Mitochondria are visible in living cells under the phase microscope. They are seen to undergo active movements during mitosis and meiosis. Mitochondria may also be seen in living cells stained with the vital dye Janus green B, which though not entirely harmless, permits the cell to remain alive for a long time so that one can observe how the mitochondria change in shape and size as the conditions vary.

Further evidence of the identity of mitochondria is provided by their movement as units during centrifugation of a whole cell and by their discreteness after cell fragmentation. They may be separated from other particles in fragmented cells by washing in salt solutions followed by differential centrifugation. In suspension they maintain their identity if adequate conditions are provided. Such suspensions of mitochondria have been subjected to biochemical analyses, and many enzymes concerned with aerobic respiration, energy transfer and storage have been identified in them. Perhaps for this reason mitochondria are concentrated in regions of greatest cellular activity: at the secretory surface of a glandular cell, near the nodes of a nerve cell at which propagation is believed to occur and in active muscle fibers. In these locations mitochondria might supply most readily the high energy compounds needed for the activities of the cells.

When cells divide, the mitochondria have been seen to orient themselves relatively symmetrically on each side of the division figure, partition between the two cells being fairly equal. In some cases mitochondria have been seen to fragment after division. It is possible that the fragments then increase in size, thus reconstituting the amount of mitochondrial material needed in each daughter cell. However, it is thought that mitochondria may also develop from minute bodies in the cytoplasm. Mitochondria have been shown to contain deoxyribonucleic acid different from that found in the nucleus, which suggests that they may be self-duplicating systems.

### *The Golgi Complex*

In the living cell under favorable conditions the Golgi complex may be stained with methylene blue and observed for a limited time. It is also visible by means of phase microscopy. It seems to be closely connected to the endoplasmic reticulum but its outlines are smooth (ribosomes being lacking) and it is chemically somewhat different from the reticulum, containing much more lipid. It can be separated from other constituents of the cell by differential centrifugation. Some question exists whether the Golgi complex really constitutes a true organelle. Cytologists are inclined to look upon the Golgi complex as an assembly area for different kinds of secretions, especially lipids, rather than as a structural unit. Into the Golgi area flow high energy compounds from the mitochondria and various secretions, probably including nucleic acids, from the nucleus.

### *Lysosomes*

By gently centrifuging certain kinds of cells lysosomes have been separated from other cell organelles and their enzymes studied, thus showing them to be true cell organelles.

### *Plastids*

The identity of chloroplasts as discrete structures has not been doubted because they are so readily seen in living cells of green plants. They have been separated by maceration of cells and fractional separation of the constituents. The internal structure of chloroplasts has also been studied by electron microscopy.

*Chloroplasts* have been found to contain deoxyribonucleic acid (DNA) different from that of the nucleus. Evidence has also accumulated that chloroplasts are self-replicating organelles, probably as a consequence of their extranuclear DNA.

Other plastids, however, have been less thoroughly studied, perhaps because their functions in plant cells are not as clearly evident as those of the chloroplast. Leucoplasts, within which starch grains are formed, are organised bodies in the cytoplasm and they contain enzymes which synthesize starch from the glucose formed during photosynthesis. Perhaps other synthetic functions are also localised in the plastids of plant cells.

### *Microtubules and Micro filaments*

Recently electron microscopists have discovered microtubules and microfilaments in the ground substance between the cisternae (larger membrane-bound cavities) and tubules of the endoplasmic reticulum. Because the microtubules occur in abundance in cells maintaining an odd shape, as in the case of the vertebrate red cell, they are thought to be structural elements. But they also appear to be important in cell division. Microfilaments occur in motile cells such as the plasmodium of the slime mold and may function in movement.

***Miscellaneous Cell Inclusions***

Occasionally oil droplets and oil globules are seen in the protoplasm, particularly in cells in tissue culture, in marine eggs and in marine protozoans known as radiolarians. These oil droplets are probably stores of nutrient or in some cases devices for ducing the specific gravity of the cell, thus making floating easier. In some cells ycogen granules occur, presumably repsenting stores of nutrient for the cell.

At present, workers have accepted the ality of the major cell organelles and are now concerned mainly with determination of their chemical constitution, emicaliemical organisation and function.

## VARIATION IN ORGANISATION OF ANIMAL CELLS

In multicellular organisms, all of the cells are not exactly alike. They have become differentiated and adapted to perform some particular function of the organism.

A *tissue* is a mass of similar cells, usually continuous, held together in a supporting matrix (secreted by the cells), performing a common function and usually forming a part of an *organ*. Intercellular fluid is almost always found in tissues. This is to be expected since every cell must have water frontage from which it obtains supplies and into which it pours its wastes. But because this intercellular fluid is not easily removed or modified, both plant and animal tissues, when used in experiments cannot be considered the equivalent of suspensions of cells.

Animal tissues are usually classified as epithelial, connective, muscular, nervous and blood. Epithelial cells are generally closely packed, like bricks in a pavement, and form solid protective layers on the outside. They line external surfaces, cavities and tubules, produce secretions and proliferate cells. Epithelial cells may be isodiametric (having equal diameters in all directions), flattened, or elongated and columnar (often the columnar cells are glandular). From the epithelial cells of glands exude various secretions. Germinal cells, which give rise to gametes, spring from the epithelial cells of gonads. In epithelial cells are found the typical organelles expected in a cell. In glandular epithelial cells the mitochondria and Golgi complex are likely to be particularly well developed. Although epithelial tissue consists primarily of cells, the external epithelium may secrete a cuticle, as in the earthworm, or a skeleton, as in arthropods.

Connective tissue, in contrast to epithelial tissue, is to a large extent composed of extracellular material and may contain relatively few cells. Some examples of connective tissues are cartilage, tendon, bone and adipose or fatty tissue.

During early development of the embryo the connective tissue cells lie close together, but during later development they secrete a gelatinous material (matrix) which separates the cells. In this gelatinous matrix are laid the fibers of.connective tissue and the salts of bone. The chemical composition of connective tissue in tendon is diagrammatically represented in Figure elsewhere in this chapter.

In muscular tissue (smooth, cardiac or skeletal) the cells are considerably modified for the function of contraction. They are elongated cells and in all cases contain contractile fibrils, these being characteristic of muscle tissue and of no other tissue. The cells of the smooth or visceral musculature of vertebrates are spindle-shaped units containing a single nucleus in the center and have fine elongate fibrillae which represent the contractile elements.

The skeletal musculature of vertebrates is highly differentiated for its function. Its unit of organisation is the muscle fiber, consisting of the confluent bodies of numerous cells, each of which is indicated by a spindleshaped nucleus. Elongate striated fibrils pack the cytoplasm of the skeletal muscle fiber, and large mitochondria are present; they are especially conspicuous in the highly active flight muscles of vertebrates and insects. In the cardiac muscle of vertebrates, striated fibers are found and the unit is modified in organisation from that of skeletal muscle. Muscular tissues of invertebrates are very diverse, but in all cases contractile fibrils are found.

The cells of nervous tissue are highly specialised for conduction of nervous impulses. From the nerve cell body containing the nucleus extend the long fibers called *axons* and the short fibers called *dendrites*. The nerve cell with all of its processes is called *a neuron. A* well developed Golgi complex appears in the cell body of a neuron, and mitochondria occur chiefly at the synapses.

In the fibers are found granular deposits of nucleic acids which probably serve as nutrient in the activities of the nerve cell. The axon of a vertebrate nerve cell is called a *myelinated fiber* when it is covered with an envelope consisting of many layers of apposed cell membranes and enclosed by sheath cells. The myelin sheath may add

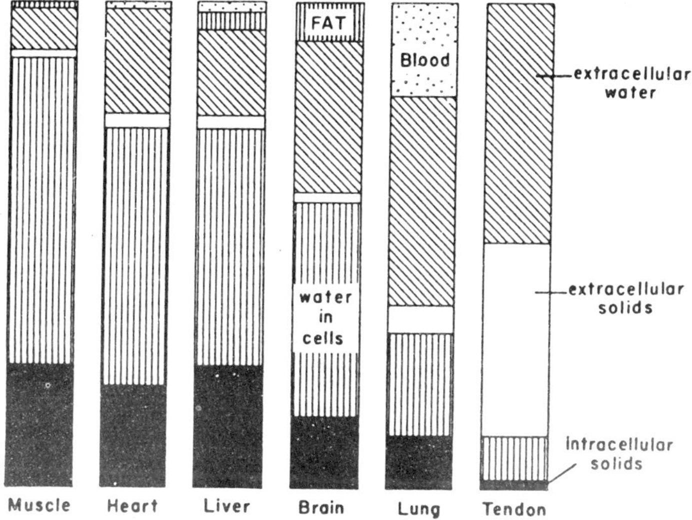


*Figure 2.4: Extracellular material in animal tissues.*

much bulk to the vertebrate nervous system and, because of it, nervous tissue like the brain contains a considerable amount of extracellular material. However, the outgoing fibers from autonomic nerve ganglia located outside the cord are not myelinated. Invertebrate nerve fibers have little if any myelin.

Blood also has cells that are highly specialised. For instance, in vertebrate blood the red cells (erythrocytes) are mainly composed of the oxygen-carrying pigment, hemoglobin. In mammals the mature erythrocytes even lose their nuclei and have none of the characteristic cell organelles. The white cells of vertebrates and invertebrates-both granular ameboid cells called leukocytes, and the more or less spherical clear lymphocytes-are much more typical cells than the red cells since they have nuclei and the other cell organelles such as the mitochondria and the Golgi complex.

## VARIATION IN ORGANISATION OF PLANT CELLS

Unlike animal cells, plant cells usually possess plastids and vacuoles and are covered by a cell wall. Mitochondria, ribosomes and a nucleus are present as in animal cells. Plant cells have structures called *dictyosomes* which are similar to the Golgi complex of animal cells.

Little agreement exists on classification of plant tissues, but the following might be cited as one possible classification: epidermal, undifferentiated, vascular (xylem and phloem), sclerenchyma and supporting or bast fibers.

The cells of epidermal tissues of plants line external surfaces of plant structures. Embryonic cells of plants, e.g., the *cambium,* are made up of cells similar in appearance and function to epithelial cells of animals. Undifferentiated plant cells, called *parenchyma,* exist in large numbers in the cortex of the stem, filling the spaces between conducting or supporting tissues.

They also make up the mesophyll and the palisade cell layer of the leaf, in which photosynthesis occurs. Vascular tissue of plants consists of the *phloem* and the *xylem.* The phloem is constituted of elongated living cells which abut upon one another. It distributes the manufactured food throughout the entire plant. The xylem consists of tubular remains of cells; it distributes water and salts to the cells of the plant. In the adult plant the xylem ducts are entirely extracellular, the cellular material having disappeared.

Cells with thickened walls, like the gritty stone cells in the pear, are representatives of *sclerenchyma* tissue. Supporting "cells" or bast fibers, consisting largely of secretions of cellulose and lignin, occur along the vascular tissues and contribute to the skeleton of the plant. Supporting fibers in the adult plant are entirely extracellular.

As a result, an organ that contains an abundance of ducts and supporting fibers, like a mature stem, has much extracellular material. In contrast, a bud or a leaf is made up largely of relatively undifferentiated cells, and, except for the cell walls, consists of living cells. Unfortunately, no convenient summary of the intracellular and extracellular constituents of plant tissue, comparable to that for animal tissues shown in Figure elsewhere in this chapter, is available, although it would be desirable since plant tissues are extensively used in cellular research.

## VARIATION IN ORGANISATION OF BACTERIAL CELLS

Bacteria and blue-green algae, unlike all kinds of true cells of plants and animals, lack a distinct nucleus enclosed by a nuclear envelope, the nuclear material being scattered in the cytoplasm. The blue-green algae differ from bacteria in that they contain chlorophyll and a characteristic auxiliary pigment and differ from most bacteria in being photosynthetic.

Plant and animal cells with true nuclei are called *eukaryotic cells* and bacteria and cells of blue-green algae are called *prokaryotic cells*. Also, prokaryotic cells either lack all the other cell organelles (so characteristic of eukaryotic cells) or possess them in a primitive, undifferentiated condition. Because bacteria have been extensively investigated, they alone are considered here, but much of what is said is applicable to blue-green algae as well.

Bacterial cells are covered with a protective *cell wall* which usually consists of carbohydrates and polypeptides but may contain protein and lipids. The bacterial cell can be made to retract from this wall if it is placed in a concentrated solution of sugar or salt, so that the cell membrane (plasma membrane) now forms the outer boundary of the living material of the cell called the *protoplast*.

The cell wall may be digested with the enzyme lysozyme, exposing the protoplast which can now be made to swell or shrink like any other cell by decreasing or increasing the concentration of salts dissolved in the medium. The cell wall evidently plays some important role in addition to protection, since protoplasts divested of the cell walls do not divide or form new cell walls.

Some bacteria, e.g., *Bacillus megatherium,* called gram-positive bacteria, have ribonucleic acid present in the surface which takes up a stain made up of crystal violet and iodine. Bacteria which are gram-negative (e.g., *Escherichia coil,* the colon bacterium) differ in possessing cell walls heavily impregnated with lipids. Thick-walled resting stages called *spores* which resist high temperature and other unfavorable conditions are formed by some bacteria (bacilli).

Bacteria lack mitochondria and the Golgi complex although they generally are filled with bodies resembling the ribosomes of true cells. As in eukaryotic cells the ribosomes may cluster to form polyribosomes during the synthesis of proteins in bacterial cell homogenates.

The plasma membrane, which is rich in oxidative enzymes, probably performs the function usually carried out by the mitochondria of plant and animal cells. It is of interest that the plasma membrane of the large bacterium *Thiovulvum majus* sinks deep into the cytoplasm, forming a multilayered structure called the *desmosome*. In the tubercle bacillus, concentric membranes, suggesting incipient mitochondria, have been discovered in the cytoplasm with the aid of electron microscopy.

Many bacteria have been shown to possess large granules (round bodies) which grow and divide and are stained with the same dyes which affect chromosomes of true cells. These bodies appear to condense before division and become more diffuse thereafter.

Rickettsias, which cause Rocky Mountain spotted fever, endemic typhus and murine typhus, are of about the size of small bacteria and in general their organisation clearly places them among the bacteria. However, to some workers they appear to be escaped mitochondria and to others, a stage between viruses and bacteria. Like bacteria, they possess a number of respiratory enzymes which are lacking in viruses. They have not yet been cultured except in cells of a host they infect.

## THE PPLO

The smallest bacteria, and the smallest cells known, are found in the group of organisms known as the pleuropneumonialike organisms, also known as the PPLO (Mycoplasmataceae). *Mycoplasma laidlawii,* a free-living strain growing in sewage, is the smallest cell so far studied. It shows a definite life cycle. In culture it exists in bodies of three sizes, minute elemental bodies 0.1 μ in diameter, intermediate size bodies and full-size cells about 1 p. in diameter.

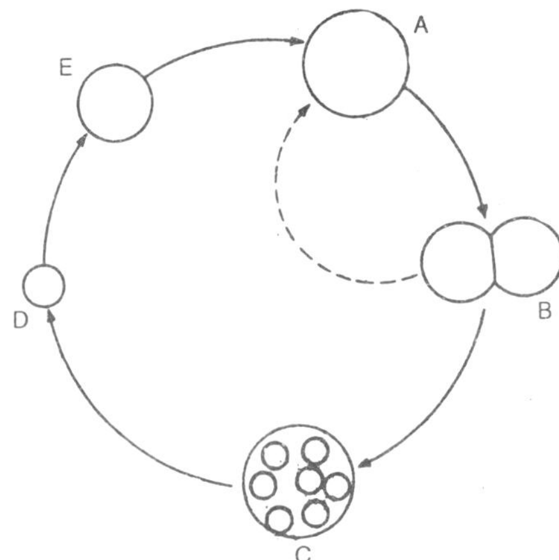


*Figure 2.5: Stages in the life cycle of Mycopla.sma laidlawii. The cell at A is about 1 μ in diameter. It may divide (B) to form two cells like itself or microbodies (C) which are about 0.1 s in diameter (D). The microbodies grow (E) to reform the full-sized cell (A).*

Studies indicate that the small bodies develop into the large cells, which can also divide to form more of their kind. *Mycoplasma gallisepticum,* which causes chronic respiratory disease in poultry, is somewhat larger (about 0.25 p) than *M. laidlawii* and has been especially useful in biochemical studies. Some of the chemical constituents are similar to those found in bacteria; others are like those found in animal cells. Of the nucleic acids found, some are linear and double-stranded (deoxyribonucleic acid), and others are in the form of spheroids combined with protein and resembling the ribosomes of other cells. Among the proteins are some 40 different kinds of enzymes, including those that catalyse the oxidation of pyruvic acid to carbon dioxide and water, reactions that liberate much energy and are found in a wide variety of cells, bacterial, plant and animal.

The lipids found include cholesterol and cholesterol esters, both typical of animal cells. The plasma membrane is flexible and electron micrographs disclose it to be a "unit membrane" similar in dimensions to the unit membrane of cells in general.

## ORGANISATION OF VIRUSES

Some viruses are larger than bacteria; for example, the psittacosis virus measures 0.75 p in diameter whereas a pleuropneumonia bacterium measures only 0.25 ft in diameter. However, most virus particles are smaller than bacteria; the virus causing hoof and mouth disease is estimated to be only 0.01 p in diameter. The distinguishing character of viruses is that they are invariably intracellular parasites of specific hosts and none are known to be free living.

Viruses attack bacteria, plant and animal cells. This is probably a consequence of their lack of energy-yielding and synthetic enzyme systems. The only enzymes demonstrated in viruses enable the bacterial virus to enter a bacterium, presumably by digesting the cell wall. Most viruses consist of a protein coat over a nucleic acid core, but lipids and lipoproteins have been claimed from some animal viruses. Viruses contain only one kind of nucleic acid, ribonucleic acid in plant and animal viruses, and deoxyribonucleic acid in bacterial viruses, whereas animal and plant cells have both.

In general, animal viruses appear to be of a more complex nature than viruses of plants and bacteria (bacteriophages), but less is known of them because they present so many experimental difficulties. All viruses lack cell membranes delimiting them from the medium. They also lack the other organelles characteristic of bacteria, including cell walls, and thus viruses represent an entirely different category of organisation. Most extensively studied are the viruses which attack bacteria and are called *phages,* or more properly, bacteriophages (bacteria-eaters as the name implies).

Only the nucleic acid portion of a phage enters (infects) a bacterium, as can be shown by radioactive markers on the nucleic acid ($^{32}P$) and on the protein ($^{35}S$). After the phage nucleic acid enters the bacterium, the suspension of bacteria and phage can be agitated violently to detach the phage particles. Infection occurs nonetheless, even though it can be demonstrated that only about 1 per cent of the labeled phage protein still adheres to the bacteria. Furthermore, nucleic acid-free "ghosts" containing all of the protein of bacteriophages are not infectious to bacteria although they attach themselves firmly.

The bacteriophage nucleic acid particle, once inside the bacterium, loses its identity (eclipse period), but after half an hour phages can again be identified (by electron microscopy). It can be demonstrated by various labeling techniques that, once inside, the nucleic acid of the phage takes command of the synthetic activities of the bacterium, suspending the normal synthetic activities of the bacterium, hydrolysing its nucleic acids and directing the synthesis of the virus type of nucleic acid and protein.

The protein coat of the virus is synthesized anew from materials taken in by the bacterial cell. About 10 minutes after infection, packaging of the virus nucleic acid in protein coats begins (as shown by electronmicrography in disrupted virus-infected bacteria) and continues for another 20 minutes, when the process is complete. At this time the bacterium is ruptured (lysed), liberating between 40 and 100 or more complete phage particles.

The effect of a bacteriophage in a suspension of bacteria *(in vitro)* plated on nutrient medium can be detected by spots of lysis-clear areas produced by liquefaction of the layer of bacteria. This occurs only after a number of life cycles of the phage.

The number of particles or the *titer* in the initial suspension can thus be assayed by the number of lysed spots.

Phages may affect bacteria in either of two ways-they may grow and lyse the bacteria as just described, (phages that cause lysis are called . virulent or *lytic phages),* or, they may become incorporated into the genetic material of the bacterium and remain in an inactive "masked" form.

That they have altered the hereditary characteristics of the bacteria is shown by the resistance (immunity) of the bacteria so infected to attacks by homologous phages which would otherwise lyse these bacteria. Phages that "hide" in this manner, dividing only when the bacterium divides, are spoken of as temperate or *lysogenic phages (i.e.,* giving rise to lytic phages). An occasional masked lysogenic phage particle multiplies, perhaps because its host has become weakened and no longer holds it in check .

Damaging the bacteria infected with lysogenic phages by radiations or other unfavourable agents leads to a rapid development of the phage into an active lytic form. The particles are then released in infection: form. Lysis of an occasional bacterium in an undamaged population can spread the infection to many previously uninfected members of the population, in this way maintaining a mild infection. Lysogeny, as the phenomenon is called, indicates the intimate relationship formed by the phage and the bacterium, a conception having great influence on microbial genetics.

## STAINING TECHNIQUE FOR LIVING AND FIXED CELLS

*Vital dyes* penetrate living cells and colour certain structures without seriously injuring the cells. For example, neutral red may be used to stain the cytoplasm, although in some cells it accumulates in vacuoles instead. Janus green B stains mitochondria selectively. Methylene blue selectively stains the Golgi complex.

Vital dyes are not entirely harmless, but they kill only after cells have been exposed to them for a long time, thus allowing time for study. Sometimes the poisonous constituent (e.g., the heavy metals in Janus green B) may be removed from a vital dye solution, making the dye much less toxic than the original sample. Although helpful, the vital dye technique has only limited use because many of the cell organelles are not stained by such dyes and, in all cases, scattering of light obscures boundaries of structures.

However, dyes have been used to demonstrate that some cell organelles such as the mitochondria, Golgi complex and vacuoles are real structures present in the living cell. Most of the early work on the nature of cell organelles was done with fixed and stained preparations. *Fixing agents,* such as formalin, alcohol, acids, salts of heavy metals or mixtures of these, precipitate the proteins and render them insoluble.

Next, water is removed from the fixed tissues *by dehydrating agents,* such as, alcohol, and the tissues are embedded in paraffin and sectioned with a microtome. The sections are affixed to slides and the paraffin is removed with xylol and washed in xylolalcohol. By washing in decreasing concentrations of alcohol the sections are partially hydrated and the proteinaceous material of the cell is then differentially stained to distinguish the structures present.

Natural dyes (e.g., hematoxylin) or basic anilin dyes (e.g., safranin and basic fuchsin) stain the nucleus selectively, and acid dyes (e.g., orange G, eosin and fast green) stain the cytoplasm. The sections are then dehydrated with alcohol. To reduce scattering of light, it is necessary to replace the alcoholic medium with a substance having the same refractive index as that of the protein particles. This is accomplished with a *clearing agent,* such as xylol, which infiltrates among the protein particles. The preparation is then mounted in balsam, which has a refractive index about equal to that of the cell proteins. As a result, it is possible to look at the cell and see clearly the stained structures that are within.

Cytochemical techniques for identification by colour reactions of cell constituents and enzymes are considered in other Chapter of this book. Some cytochemical techniques have been developed at the electron microscope level.

## AUTO RADIOGRAPHY

If a cell is suspended in a solution containing a substance labeled by a radioactive isotope, which the cell is capable of incorporating, it is often possible to determine where in the cell incorporation occurs and to what extent. For example, $^3$H-uridine is presented to cells for a brief time and the cells are mounted on a slide and fixed. A layer of photographic emulsion is deposited over the preparation.

The radioactive emission from the $^3$H-uridine will affect the silver halide in the emulsion. On development, the grains affected by the radiation (beta particles of low energy value) are reduced and appear as black grains or traces over the structure into which the $^3$H-uridine had been incorporated. If the grains appear over the nucleus we know that the nucleus incorporates uridine into ribonucleic acid (RNA). Similar experiments can be performed with $^3$H-thymidine, which is selectively incorporated into deoxyribonucleic acid (DNA), and with $^3$H-amino acids, which are incorporated into proteins. Autoradiography has been of inestimable value in localising the sites of a variety of syntheses in the cell. Recurrent reference to the use of this technique will be made in subsequent chapters.

## MICROSCOPICAL TECHNIQUES

The ordinary *light microscope* makes it possible to see stained structures so long as they are larger than the limit of resolution. The smallest object, d, resolved (seen clearly and distinctly from similar sized objects separated by like small distances) by the light microscope may be determined from the following equation:

$$d = \frac{\lambda}{\mathrm{NA+na}} \qquad (2.1)$$

In this equation A is the wavelength of light used, NA is the numerical aperture of the objective and na the numerical aperture of the condenser. Since for critical study the numerical aperture of the condenser is chosen equal to that of the objective, the equation is often written:

$$d = \lambda / 2NA \tag{2.2}$$

The numerical aperture is given by the equation:

$$NA = n \sin \alpha \tag{2.3}$$

Here n is the refractive index of the light path in the formation of the microscopical image, and α is half the angular aperture of the light beam entering the condenser. When immersion oil is used between condenser and slide, and slide and objective, the numerical aperture may be as large as 1.6, although in practice it is seldom over 1.4.

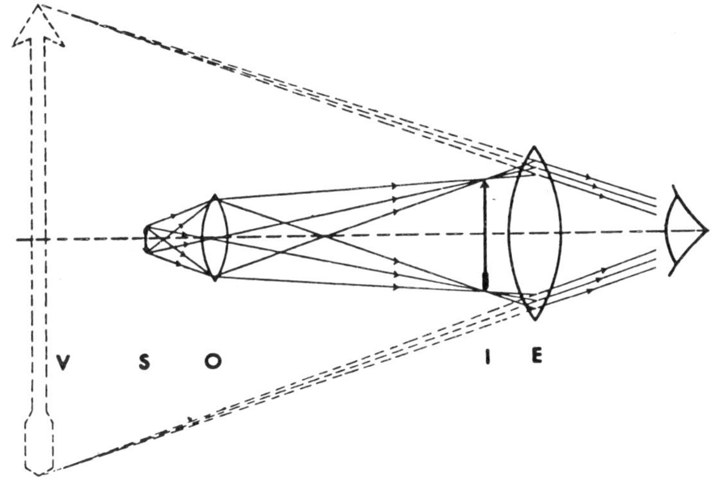


*Figure 2.6: Principle of the compound microscope. The intermediate image (I) of the subject (S) formed by the objective (0) is enlarged by the eyepiece (E). The virtual image (V) is seen by the eye. A real image of V is recorded with a camera.*

The factor that limits resolution with the light microscope is therefore the wave length of light. Under optimal conditions the smallest object seen is theoretically about one-third the wavelength of visible light or about 0.13 p in blue light.

Usually the limit is set at about 0.2 p because the eye is most sensitive in the yellowgreen part of the spectrum (0.55 p) and a maximum numerical aperture of 1.6 is seldom achieved. An object of this size is visible only if the structure shows maximum contrast. In order to increase the contrast between unstained structures and to obtain greater resolution of details of cell organelles, various types of microscopes have been developed, each of which will be considered in turn. The *dark field microscope* is an ordinary microscope with a special condenser. By use of a dark stop in the condenser the light from the center of the field is removed and the object is illuminated only by an oblique beam of light as seen in Figure elsewhere in this chapter.

The object is therefore seen by light reflected and scattered from interfaces in the object. With dark field microscopy the outlines of the cell, the nucleus, mitochondria, oil droplets, vacuoles and various inclusions may be readily identified. Moreover, in cells undergoing division, the centrosomes, asters, spindles and chromosomes may be observed. Objects smaller than those seen with the ordinary light microscope may be detected with dark field microscopy but not resolved.

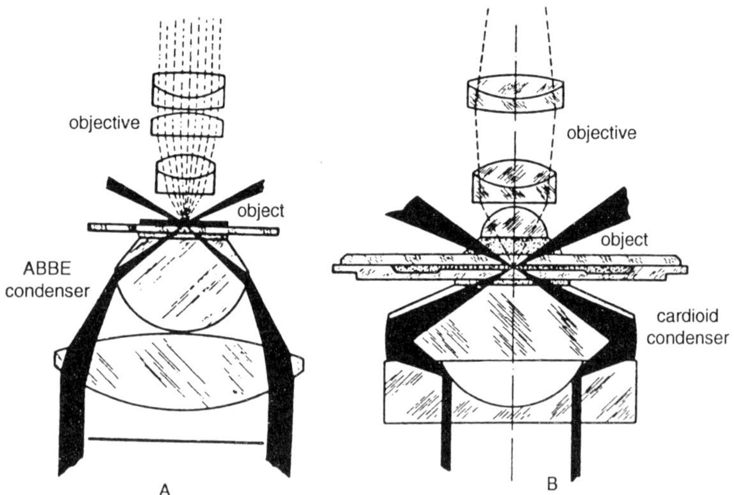


*Figure 2.7: Dark field microscopy. The dark, field stop shown in A is much less effective than the field condenser shown in B.*

The *polarising microscope* is like a light microscope except that the condenser and the ocular are equipped with polarising optics, each of which transmits only planepolarised light. The condenser is then spoken of as the *polariser* and the ocular as the *analyser*. If the optics of the condenser and ocular are crossed (at 90° to one another) no light is transmitted unless an interposed object rotates the plane of the light. The manner in which this occurs and the usefulness of the microscope in biology is briefly outlined in the following paragraphs.

Light, passing through a doubly refracting crystal such as calcite, is separated into two rays which vibrate at right angles to one another. Such light is said to be polarised. The *Nicol prism,* one of the best known polarisers, is made by cutting a rhomb of Iceland spar (a pure, transparent form of calcite) in halves along the line of one of the optical faces, polishing the cut faces and cementing them together with Canada balsam.

As seen in Figure elsewhere in this chapter, unpolarised light entering one end is separated by the prism into two beams, one vibrating in the plane of the figure, the other at right angles to it. One of these beams is totally reflected by the layer of Canada balsam and is usually absorbed by the opaque case covering part of the rhomb.

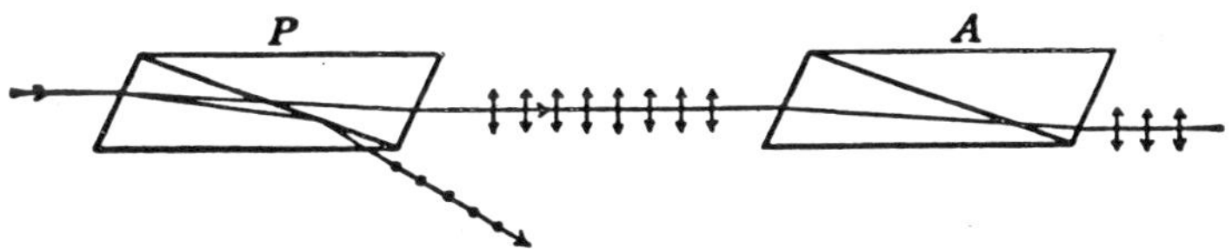


*Figure 2.8: Polarisation of light by a Nicol prism. Only the beam in the plane of the sheet has been permitted to pass; that at right angles is deflected. The second Nicol prism, if oriented like the first, has no effect on the polarised light. However, if it is rotated through 90° the beam will be extinguished.*

The other beam is transmitted. Hence, a Nicol prism transmits plane-polarised light. If a second Nicol prism is used in series with the first, and in the same position, the polarised light is transmitted. If, however, the second prism is rotated through 90° in relation to the first prism, the light fails to pass. But if an object like quartz, which polarizes light, is placed in the beam of light between the two prisms, and if the two Nicols are at right angles to each other (crossed), instead of a plain dark field, a brilliantly lit object is seen against a dark background, indicating that the object has polarised the light passing through it.

If the second prism is then rotated away from 90°, the field suddenly becomes dark. The angle required to rotate the second prism in order to extinguish the light is a measure of the degree to which the plane of polarised light has been rotated by the interposed object. The second prism is called an analyser because by moving it one can analyse the degree to which the plane of polarised light has been rotated by an object inserted in the light path.

Various natural substances are observed to be *isotropic*; that is, light passes through them equally well in all directions. Others, which affect the passage of light unequally in different directions, are *anisotropic* or *birefringent*. The ray passing directly through a crystal is called the *ordinary ray;* the ray which is deviated is called the *extraordinary ray*. If the velocity of the extraordinary ray passing through such a material is less than the velocity of the ordinary ray, the crystal is said to be negative (e.g., quartz). If the extraordinary ray has a velocity greater than the ordinary, the crystal is said to be positive. *Crystalline* or *intrinsic birefringence* is found where the bonds between molecules (or ions) have a regular asymmetrical arrangement. The birefringence in this case is independent of the refractive index of the medium.

*Dichroism* occurs when the absorption of a given wavelength of polarised light changes with the orientation of the object. For example, the degree of absorption at wavelength 260 nm. by nucleic acids in cells depends upon whether the light is parallel to, or at right angles to, the oriented purine and pyrimidine residues. Dichroism is a sensitive indicator of orientation of molecules in a structure.

However, it is seldom seen in biological objects in the visible spectrum except in some stained preparations where it results from orientation of dye molecules. It may be seen at 260 nm. in the sperm nuclei where the concentrated nucleic acids appear to be oriented.

Birefringence of inanimate objects, then, is the result of their crystalline nature. Birefringence of biological materials indicates that their molecules are oriented with respect to one another much as in a crystal. Some particles-for example, nucleoprotein molecules-line up only if the solution is flowing; what they show is therefore called *flow* or *streaming birefringence.*

This indicates that the molecules are elongate and line up in the path of least resistance to a current, just as sticks line up in a brook. Organisms are more likely to show *form birefringence,* because of regular orientation of tiny particles (rods or planes) differing in refractive index from their surrounding medium. Birefringence in such cases disappears when the material is immersed in a medium of the same refractive index.

Oriented rods have different refractive indices for directions parallel to and at right angles to the major axis. Therefore, by making observations with polarised light one may determine in which direction in a given structure the rods lie.

Polarising optics have been especially useful in study of biological membranes and of the division figure in cells. They are also useful in studying fibrillar structures.

The *phase contrast microscope,* which has the same resolving power as the ordinary light microscope, enables one to see structures in the living cell by taking advantage of the slight differences in refractive index between any two structures to improve their visibility. The phase contrast microscope does this by the use of a special diaphragm in the condenser and a phase plate in the objective. The annular diaphragm used in the condenser of the phase microscope permits light to pass through the condenser as a hollow cone, the remaining light being absorbed. This cone is focused on the object.

The phase plate placed at the back focal plane of the objective is a transparent disk containing a groove (or elevation) of such a size and shape as to coincide with the direct image of the substage annular diaphragm formed at the back focal plane when no object is viewed. If an object is placed between the condenser and the objective, in addition to the direct image a number of overlapping diffracted images of the diaphragm then appear at the back focal plane of the objective.

The depth of the groove (or the elevation) in the phase plate of the objective is so made that the two sets of rays forming the direct image and the diffracted image differ in optical path by a quarter wavelength of the illuminating beam of light. Under these conditions the phase difference, which is not seen by the eye, is converted to an intensity difference which we see.

In the bright contrast phase optical system the two sets of rays are added to make a brighter image as in Figure elsewhere in this chapter, whereas in dark contrast phase they partially cancel one another making a more contrasting darker image, as in Figure elsewhere in this chapter.

The phase contrast microscope (bright or dark contrast may be used), in which a small difference of refractive index is exaggerated, enabling one to distinguish adjacent structures, has made it possible to observe structures previously very difficult or impossible to see. The behavior of chromosomes of living cells during mitosis or meiosis can be followed with ease, and *many* of the other organelles may be studied.

The phase contrast microscope has furnished the most convincing evidence that many of the organelles identified in fixed and stained preparations of the cell are real and not artifacts. However, the images are often surrounded by a halo and the difference in intensity between two areas of unequal refractive index diminishes with the distance from their boundary, thus causing some distortion.

The *interference microscope* avoids the residual halo which accompanies the image seen under a phase microscope because the aperture of the objective of such a microscope is not limited by a phase plate and the same area of the objective is used for both interfering beams. The relative phase of the two beams is varied continuously so that any part of an object can be given maximum or minimum contrast at will.

As observed in Figure elsewhere in this chapter, a horizontal glass

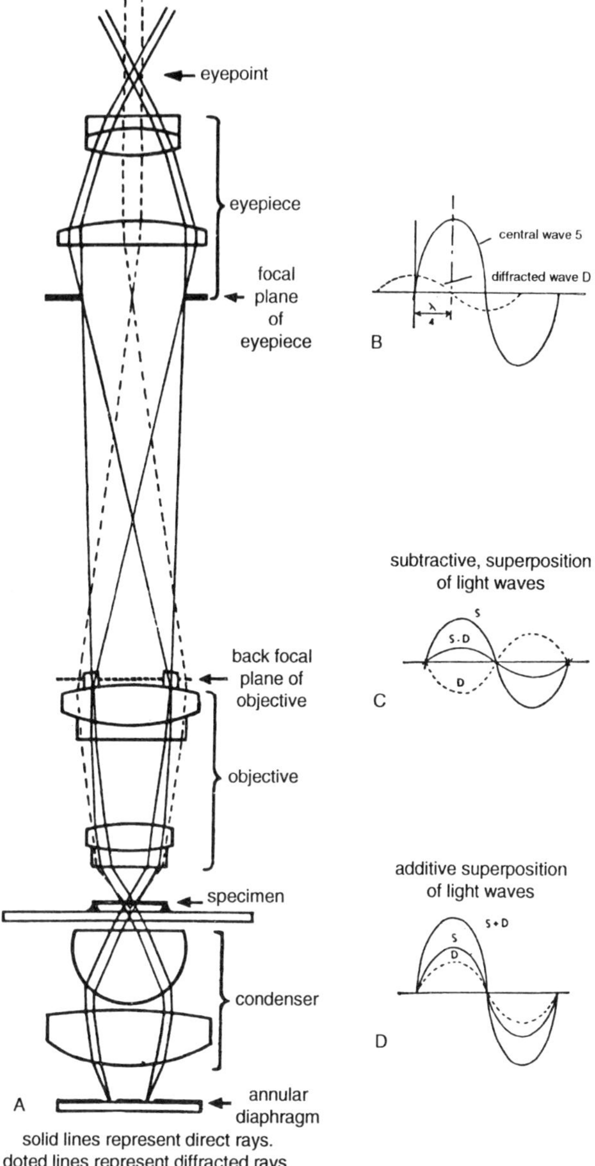


*Figure 2.9: A shows the light path in a phase microscope. B shows the normal retardation by wavelength of light diffracted by an object and its difference in phase from the light passing through the surrounding medium. By phase optics the two waves are superimposed to reinforce each other in bright contrast phase as shown in D or to subtract from each other as in dark contrast phase shown in C.*

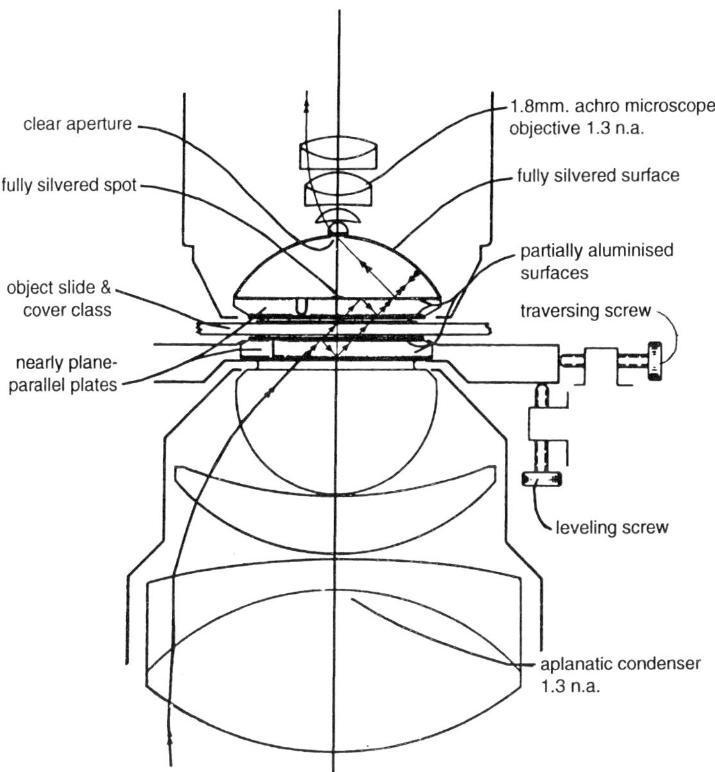


*Figure 2.10: Dyson interferometer microscope, light path.*

plate, L, nearly parallel to a similar plate, U, above the object, intercepts the illuminating cone of light from the condenser (both surtaces of this plate are partially aluminised).

A portion of the iluminating cone passes through the plate and is focused on the object, and a portion reflected downward at the upper surface plate L and is then reflected upward its lower surface. Although only one am traverses the object, two beams averse the object plane. The beam fail; to pass through the object, called the *comparison beam,* is internally reflected ice within the plate. It will be noted at it has a diameter much larger than at of the field.

A second horizontal plate, U, identil to the first and just about parallel to it, is located above the plane of the object. Both of its surfaces are also partially aluminised, but in addition its upper surface has an axial opaque spot somewhat larger in diameter than the maximum field of the microscope. This spot prevents the direct vertical light beam from the source which has passed through the object from reaching the objective, but allows the light from the beam which has traversed the object and then enters the plate to the side of the spot to reach the objective. Part of the illuminating beam which has traversed the object is reflected twice in plate U and part of the comparison beam traverses this plate without reflection; the two beams will emerge from the upper surface in coincidence when the plates are correctly aligned.

The observer above the upper plate sees two coincident fields of view, one an image of the light source only and the other an image of the light source upon which is superimposed an image of the object. Aside from the interference resulting from passage through the object, the phase difference between the two images will be zero.

The combined images may be viewed by a normal microscope objective. The interferometer plates L and U are movable so that the path difference between the two coincident fields can be varied at will over the whole area of vision. Maximum contrast cannot be obtained for objects much larger than the field of view, because the object is then traversed by a portion of the background beam.

Because an object such as a cell contains various substances, it has an optical path different from that of the surrounding water. For this reason when the background is adjusted to maximum light intensity (bright field), the cell appears darker and it is possible at a glance to note the distribution of structures in the cell much as if a stained preparation were being observed in an ordinary light microscope.

More than this, it is possible to determine the average amount of dry material (dry mass) present in a given area of the cell by the use of the interference microscope, because the greater the amount of material present in the cell, the greater will be the difference between its optical path (refractive index times thickness) and that of the surrounding fluid.

The difference in optical path is determined by the difference between interference bands in the surrounding fluid and in the object when the upper wedge plate is rotated so that the wedge axes are no longer parallel. The field is then crossed by parallel interference fringes. These fringes are seen as lighter and darker bands of the same colour when monochromatic light is used, or bands of colour on either side of white bands when white light is used. For example, one-tenth wavelength displacement for monochromatic green light (546 nm.) is equivalent to a dry mass of $10^{-13}$ gm $/\mu^2$.

The *ultraviolet microscope* looks like the ordinary light microscope but its lenses are made of quartz to permit transmission of wavelengths of invisible ultraviolet as short as 220 nm. (visible light covers the range from blue at 390 nm. to the far red at 780 nm.). The invisible ultraviolet light is then projected upon a screen which fluoresces in the visible spectrum.

Usually, however, it is recorded photographically. Because chromosomes absorb more short ultraviolet radiation than cytoplasm does, they are readily seen in photographs taken with ultraviolet light at 260 nm.. Under the ultraviolet microscope resolution of fine structure is about twice that obtained with the ordinary light microscope, as may be judged from application of equations 2.1 and 2.2, the wavelength of short ultraviolet light being about half that of the visible. Unfortunately ultraviolet radiation damages living cells, so that prolonged observation of cells so illuminated is not possible.

An instrument which may serve as a valuable adjunct to all the microscopes considered here is the *television camera attachment.* Since the electron tube used in it "sees" at low light intensities and can be made sensitive to a restricted part of the spectrum (a narrow band in the green, blue or ultraviolet), much clearer pictures can be obtained than with ordinary microscopy.

Furthermore, a prolonged illumination is avoided since the image may be viewed on a phosphorescent screen which retains the image for a minute or two after exposure. Momentary illumination with light of low intensity suffices. This attachment, therefore, is especially useful for viewing living cells under ultraviolet light.

The *electron microscope,* which was developed on the eve of World War II, has proved a powerful tool for analysis of large chemical molecules, viruses and, to some extent, cellular structures. In the electron microscope the beams of electrons are focused by magnets which serve as lenses. The object must be viewed on a fluorescent screen or photographed. The electron microscope has a resolving power 400 (theoretically 2000) times that of the light microscope. Although the wavelengths accompanying the electron beams are short (e.g., a 50,000 volt electron beam is accompanied by a wavelength of 0.05 Å), the numerical aperture of the electron microscope is only about 0.0005.

The image results from differential scattering of electrons from the molecular constituents of the cell. Scattering is greater the denser the material, regardless of its chemical composition. Even a gas scatters electrons; therefore, the biological object must be studied in a vacuum, since electron scattering from the gas molecules would obscure the differential scattering of the structures in the biological object. The preparations must be dry because the water vapor in the evacuated preparation

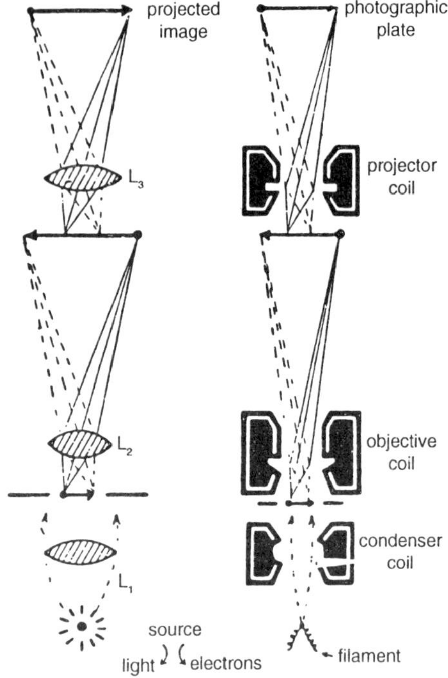


*Figure 2.11: Comparison of optical paths light and electron microscopes.*

would have the same effect as a gas. A thin slice of the object must be used; otherwise the electrons would all be scattered and the electron micrograph would reveal no detail. Since biological material varies little in density, it must be stained differentially, i.e., with some structures taking up more stain than others. *Positive staining* is obtained by fixing the cell in salts of heavy metals, e.g., osmium tetroxide or potassium permanganate.

In this case the electron-dense (i.e., opaque to electrons) metal adheres differentially to structures of the cell, causing them to scatter the electrons more than the surroundings. In other cases *negative staining* is used; that is, the object is infiltrated with a heavy metal compound such as phosphotungstic acid or uranyl acetate. These materials do not adhere to the biological structures but fill the interstices between them, making these areas electron-dense in contrast to the biological structures which are then seen as relatively electron-transparent structures .

The main disadvantages attendant upon use of the electron microscope are the required thinness of sections and the necessity of studying them dry (in a vacuum). The first difficulty has been overcome largely by the development of microtomes that can make sections 0.2 to 0.02 $\mu$ in thickness.

However, removal of water and salts dissolved in it as well as the use of stains are likely to alter structure; therefore, membranes, cytoplasm or formed structural components seen by electron microscopy must be interpreted with some caution. Notwithstanding these facts, it will become apparent in studies presented later that much has been learned about the cell by electron microscopy, and about structural elements such as the cuticle and the cell wall.

Because of the small contrast between some biological objects and the background even after staining, *shadowcasting* with heavy metals is often used, especially to bring out three-dimensional topography. Heavy metals, especially gold, platinum and chromium, are evaporated in a vacuum at an oblique angle to the object. The metal piles up like a windrow and outlines the object. Shadowing may be performed from several directions by moving the object during the process of evaporation.

When the object is too thick for shadowcasting *a replica* can be made. In this case the object, mounted on a grid containing holes, is subjected *in vacuo* to evaporation of platinum and carbon and becomes coated to the depth of several hundred angstroms. The preparation is then treated with hydroxide to remove the specimen and the replica is floated onto another grid and prepared for electron microscopic study.

*Surface spreading* has also been a useful technique for study of cell fine structure. The spread layer is "swept" in the trough to concentrate it, and mounted on a grid, dried and shadowed for electron microscopic study. In this manner a bacterial protoplast (bacterium from which the cell wall has been removed with the enzyme lysozyme) that is ruptured osmotically on the surface of a vessel filled with distilled water is found to liberate its chromosome as a single long fiber. The same technique has been used for a study of interphase chromosomes in cells of plants and animals and for a study of microtubules present in some cells.

***Freeze-etching***

The images of various cell organelles seen with the electron microscope have been criticised because they are structures seen in cells that

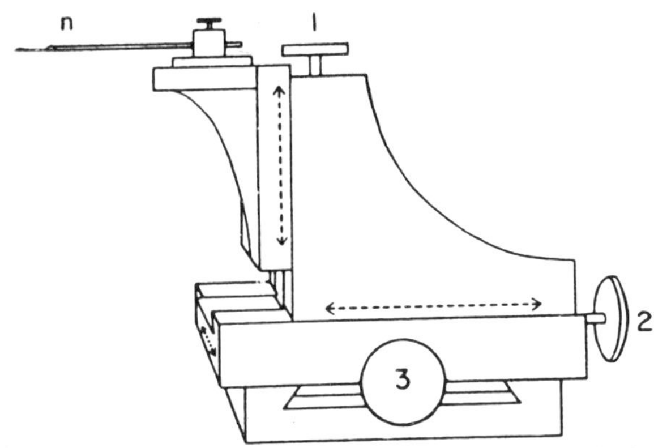


*Figure 2.12: One type of micromanipulator. It consists essentially of a microlathe enabling movements of the needle (n) in the three coordinates of space by manipulation of screws 1, 2, and 3. Another instrument (a mirror image of the one shown) stands on the other side of the microscope, all mounted on a heavy stand.*

have been fixed and stained. Are they real or are they artifacts introduced by the chemical procedures and drying?

It is disquieting also that different fixatives sometimes give different images; for example, osmium tetroxide fixation shows a plasma membrane consisting of a single line whereas permanganate fixation shows the plasma membrane consisting of two outer electron-dense layers and an inner less electron-dense layer.

It has proved possible to study the cell organelles by the use of unfixed cells for electron microscopy, in the method called *freeze-etching*. The object to be studied is placed in 20 per cent glycerol and is frozen at - 100° C. It is mounted on a chilled holder and splintered with a knife along natural cleavage planes, usually along surfaces of membranes. Occasionally the structure is cross-fractured, giving cross sections of organelles.

The splintered preparation is freeze-dried and covered with a platinum and carbon coating in the high vacuum of a freezing ultramicrotome. When the preparation is dry the vacuum is broken and the preparation is placed in water to float the replica off the carrier. The replica is then washed in basic solution to remove the cellular material and the replica is mounted on a grid and dried for electron microscopy.

Such replicas give the outlines of various cell structures and verify the structures seen in thin fixed and stained sections of the mitochondria, smooth and rough endoplasmic reticulum, double nuclear envelope with pores and plugs, unit plasma membrane of the cell,

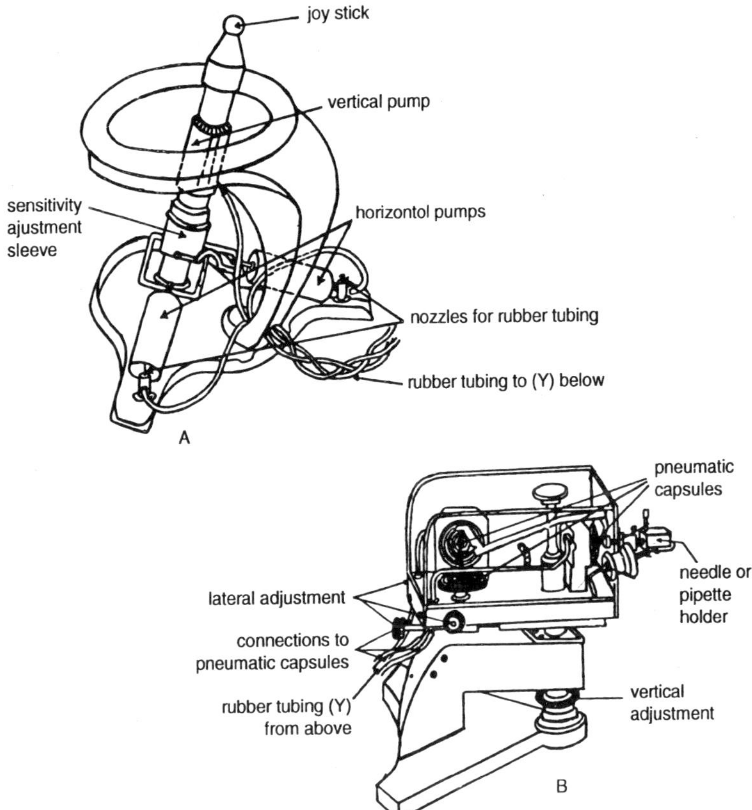


*Figure 2.13: The De Fonbrunne pneumatic micromanipulator. A movement of the control or joy stick is duplicated by the needle (or other instrument) with a reduction in magnitude of 1/50 to 1/2,500, set at the sensitivity adjustment sleeve.*

chloroplasts and so forth. The views of structures are somewhat three-dimensional. Since some cells (e.g.., yeast) subjected to these freezing and drying procedures revive and show no evident change in properties, it is assumed that electron micrographs of cells so prepared give images of cell structures in as nearly normal condition as is at present possible.

The fact that the structures compare favorably in size and appearance to those seen in fixed and stained cells is, in some measure, confirmation of the validity of the conventional method using fixatives.

## MICROCINEMATOGRAPHY

In many cases considerable information has been obtained by time-lapse photomicrography. Details that escape the observer on a single viewing become apparent after viewing the record repeatedly. Familiar examples are mitosis and meiosis in cells and irradiation of cells with a microbeam. The components used in such studies are familiar; the details must be sought elsewhere.

## MICROMANIPULATORS

Since the end of the last century micromanipulators of one kind or another have been in use, first for handling single cells and later for microsurgery upon them. The pneumatic micromanipulator of De Fonbrunne has had wide use for microsurgery because of its ease of operation, speed and sensitivity of response. Essentially it consists of three pumps, two horizontal and one vertical, which connect by way of rubber tubes to the pneumatic capsules of the instrument holder (for needle or pipette). The movements in the horizontal plane of a single lever or joy stick are reflected at a ratio of anywhere from 1 :50 to 1:2500, according to the setting decided upon. The vertical plane level is set by the screw on the lever.

An electric micromanipulator utilising thermal expansion wires has been described but has not had much use. Production of microforges for the controlled manufacture of microtools has also facilitated microdissection. Important information on the function of the nucleus of cells has been gained by the use of such tools for transplantation of nuclei between cells. The micromanipulator thus makes the microscope more than an observational tool, and properties of many of the cell organelles have been explored by its use. Even bacteria have been dissected under high power.

# 3

# PLASMA MEMBRANE

Eukaryotic cells are organised around systems of membranes. The plasma membrane envelops the cell and separates it from its surroundings; the membranes of internal organelles divide the interior into compartments with distinct biochemical environments and functions. Hydrophobic regions within membranes also provide nonaqueous environments in which certain reaction systems, such as electron transport, take place exclusively.

Realisation of the fundamental importance of membranes to cell structure developed gradually in the 1940s and 1950s. Biochemical investigations at this time revealed that some reaction systems of mitochrondria and chloroplasts could proceed only if the organelles or membranous fractions of the organelle membranes were maintained in intact form. Organisation of the enzymes of these organelles on or within membranes greatly modified their catalytic properties ; in some cases, dissolution of membrane structure abolished enzymatic activity completely.

These biochemical indications of the importance of membranes in cells were complemented in the late 1950s and the 1960s by the results of morphological studies with the electron microscope, which revealed the abundance of membranes in cell structure.

Interest and investigation in recent years have centered on the molecular structure and function of membranes. This research, which has involved the combined efforts of scientists in disciplines as varied as biochemistry, biophysics, physical chemistry, immunology, and cell biology, has revealed much new information about the liquid and protein components of membranes and their organisation in membrane structure.

## PLASMALOGENS

Some membrane phospholipids have structures differing from the phosphoglycerides to some degree. The plasmalogens, also called *ether lipids* are simile to the basic phosphoglyceride structural plan except that the hydrocarbon chain at the 1-carbon is attached by an ether instead of an ester linkage. The hydrocarbon chain at the 2. carbon of the backbone is attached to the usual ester linkage, and the 3-carbon carries a phosphate group and alcohol as in the phosphoglycerides. In the plasmalogens, the alcohol most commonly found in this position ethanolamine. Plasmalogens are commonly found in nerve and muscle membranes.

### Sphingolipids

The sphingolipids form a class of membrane lipids superficially similar in structure to the phosphoglycerides. The sphingolipids differ in the chemical group forming the backbone of the molecule, which consists of a long-chain, nitrogen-containing alcohol called *sphingosine.* Sphingosine contains two chemical groups that interact with other substances to form membrane sphingolipids, an amino (-NH,) and a hydroxyl (- OH) group.

In membrane sphingolipids, a long-chain hydrocarbon tail derived from a fatty acid is attached by an amide linkage to the amino group. The hydroxyl group may be unbound, forming a class of membrane sphingolipids called *ceramides*. More commonly, the hydroxyl site is linked to a complex carbohydrate group or by a phosphate group to one of a variety of polar or charged substances, including alcohols such as choline or ethanolamine. Figure shows sphingomyelin, a phosphate containing membrane sphingolipid of this type. The hydrocarbon chains at one end of the molecule and the polar groups at the other end also give the sphingolipids a strongly amphiuathic character.

Sphingolipids with carbohydrate groups attached at the polar ends are common as the major membrane glycolipids of both plant and animal cells. They are especially abundant in plant cell membranes and in the membranes of nerve and brain cells. In mammals, practically all of the membrane glycolipids are sphingolipids.

These mammalian glycolipids are restricted almost entirely to plasma membranes, where they orient with their polar carbohydrate groups extending from the outer membrane surface. Their function in this location has been linked to cell-to-cell recognition and adhesion. Some glycolipids at the cell surface may also act as receptors that bind to specific substances taken up by cells. Deficiencies inn the metabolism of glycolipids are the cause of several important human disorders, including Tay-Sachs disease, in which normal breakdown and turnover of carbohydrate-containing sphingoli pids is deficient. The sphingolipids that accumulate as a result interfere with nerve and brain functions, leading to paralysis and severe mental impairment.

The carbohydrate groups attached to membrane sphingolipids usually contain six-carbon sugars such as glucose, galactose, mannose, fucose, or one of the various forms of sialic acid. These may be linked singly or in straight or branched chains containing as many as fifteen sugar residues. Because of the different ways in which the sugar units may combine and link together at the polar end of a sphingolipid, these molecules can have an almost endless variety of possible structures. The same sugar residues also link together to form the carbohydrate groups of membrane glycoproteins.

**Sterols**

*Sterols* based on a framework of four carbon rings, are found in both plant and animal membranes. In contrast to the phospholipids and glycolipids, sterols are almost completely nonpolar substances. Polar groups are limited to a single hydroxyl at one end of the molecule. Ar a result, the sterols are only slightly amphipathic.

The predominant sterol of animal membranes is *cholesterol.* Plasma membranes in animals contain a significant quantity of cholesterol, amounting to as much as 30% of the total membrane lipids by weight. Membranes of the cell interior contain reduced amounts, ranging from as little as 2-4% of the total membrane lipids in mitochondria to 10% in membranes of the endoplasmic reticulum. Plant membranes contain cholesterol in small amounts and larger quantities of related sterols called phytosterols.

The role of cholesterol in membranes remains uncertain. The interior fluidity of membranes is altered when cholesterol is present, and cholesterol also reduces the permeability of membranes to water and other polar substances. Whatever its functions, cholesterol is not likely to form an essential part of the basic framework of membranes since, except for plasma membranes, must cellular membranes contain relatively little cholesterol. Sterols of any kind are rare or completely absent in bacterial and blue-green algal membranes.

## MEMBRANE PROTEINS

Most membrane proteins are functional units that reflect the biological activities of membranes. Each type of cellular membrane thus has its own characteristic assemblage of proteins, in amounts that reflect the activity of the membrane in enzymatic catalysis, cell-to-cell recognition and adhesion, and transport.

Proteins isolated from most membranes separate into 20 to 30 major bands on SDS electrophoretic gels with molecular weights ranging from about 12,000 to 250,000. These major bands probably represent the lower limit in numbers of proteins actually present, because many additional proteins probably occur in amounts too small to be detected as distinct bands.

This is true of many enzymes, receptors, and transport proteins. Others comigrate ; that is, they are so similar in molecular weight that they form a single composite band. Therefore, most membranes likely contain many more species of proteins than the 20 to 30 polypeptides usually separated on gels.

One of the surprising observations about the membrane proteins isolated to date is that their content of hydrophobic amino acids is, in most cases, not significantly higher that the hydrophobic portion found in the soluble, nonmembrane proteins of the cytoplasm. This was demonstrated nicely by R.A. Capaldi, who compared the relative proportions of hydrophobic and hydrophilic amino acids in several hundred cytoplasmic and membrane proteins.

Capaldi found that, with few exceptions, the insoluble membrane proteins contain approximately the same proportion of hydrophobic amino acids as the soluble cytoplasmic, nonmembrane proteins-slightly less than 50%.

**Table 3.1: Percent Cholesterol of Total Lipids in Biological Membranes (in Average Values).**

| *Membrane Type* | *Cholesterol* |
|---|---|
| Erythrocyte plasma membrane | 45 |
| Myelin | 38 |
| Liver cell plasma membran | 26 |
| Endoplasmic reticulum | 10 |
| Mitochondrial membranes | 2–4 |

The hydrophobic character of membrane proteins depends instead on either their folding conformation, which places hydrophobic amino acids on the exterior and hydrophilic ones on the interior, or on segregation of hydrophobic and hydrophilic residues in different parts of the proteins' amino acid sequence. This folding or segregation produces a protein with amphipathic properties similar to those of the membrane lipids, that is, with distinctly polar and nonpolar ends or regions. For example, in the protein *glycophorin,* a major constituent of vertebrate red blood cell membranes, the middle portion of the amino acid sequence contains a stretch of more than 30 hydrophobic residues. The molecule folds so that the hydrophobic central portion of the amino acid chain is maintained as a separate, strongly hydrophobic region of the protein. Other more hydrophilic parts of the sequence also fold into separate hydrophilic domains, giving glycophorin a strongly amphipathic character.

## GLYCOPROTEINS

Glycophorin also illustrates the typical form taken by membrane glycoproteins, the proteins with attached carbohydrate groups. One end of the glycophorin polypeptide chain carries complex sugar groups in branched and straight chains. The carbohydrate chains in glycophorin include fucose, galactose, mannose, glucosamine, and sialic acid. Other membrane glycoproteins contain straight or unbranched chains of the same and other carbohydrates in similar arrangements, in mixed groups containing from 4 to 15 residues.

In general, the same sugars occur in both membrane glycoproteins and glycolipids. The carbohydrate groups of glycoproteins and glycolipids occur almost exclusively on the outer surface of the plasma membrane, which gives the cell surface what is often described as a "*sugar coating*". More technically, the ceil surface layer of carbohydrates is called the *glycocalyx* (the prefix *glyco*-is derived from the Greek *glykys,* meaning "*sweet*"). Of the total plasma membrane lipids and proteins, usually no more than 10% by weight consists of glycolipids and glycoproteins. In animals, most of the glycocalyx is provided by glycoproteins ; in plants, by glycolipids. Among the various important functions of these substances in animals are those already mentioned : cell-to-cell recognition, cell adhesion, and binding specific substances taken up by cells.

### Phospholipid Suspensions in Water : Bilayers

When placed in water, phospholipids such as the phosphoglycerides readily disperse into suspensions called *bilayers.* Within a bilayer, individual phospholipid molecules are packed in parallel at right angles to the plane of the bilayer, in an orientation that satisfies their amphipatbic properties. The polar phosphate groups face the surrounding aqueous medium, and the nonpolar hydrocarbon chains associate end to end in the membrane interior in a nonpolar region that excludes water.

### Physical Properties

The bilayers formed by pure phospholipid molecules suspended in water share many properties with cellular membranes. One of the most interesting is the *phase transition,* a rearrangement of hydrocarbon chains within bilayers brought about by changes in temperature.

At low temperatures, the hydrocarbon chains are tightly packed and their motion is restricted This tightly packed site is termed the *gel* or *crystalline* phase. As the temperature increases, both the space separating adjacent phospholipid molecules and the motion of hydrocarbon chains increase abruptly.

At this temperature, the phospholipid bilayer "melts," and the hydrocarbon interior becomes fluid. These changes can be followed by X-ray diffraction and other techniques, which indicate that the increased freedom of the phospholipids above the phase transition includes rotation of entire molecules around their long axis, flexing of hydrocarbon chains, and lateral movement or diffusion of individual molecules through the bilayer.

The phase transition can be precisely followed by a techni. que called *differential calorimetry,* which measures the amount of heat required to maintain a constant rate of temperature change in a bilayer. The phase transition is marked by a peak in the amount of heat absorbed by a bilayer, since the change from gel to fluid state is endothermic (absorbing rather than releasing energy).

The peak temperature for a given phospholipid depends on the length and degree of saturation of the hydrocarbon chains and the type of polar group present in the phospholipid. Generally, the longer the hydrocarbon chains, the higher the phase transition temperature. Double bonds, since they introduce "kinks" or bends in the hydrocarbon chains that interfere with tight packing, allow the bilayer to remain fluid at lower temperatures and thus reduce the melting temperature.

Measurements using differential calorimetry show that bilayers made from a single type of pbospholipid undergo the phase transition at a characteristic, sharply defined temperature. Mixtures of phospholipids broaden the temperature range of the transition phase considerably.

Biological membranes undergo a similar phase transition, which can be detected by differential calorimetry as in artificial bilayers. The temperature range of such phase transitions is broad, indicating that mixtures of phospholipids rather than single types are present. The melting temperatures are such that the phospholipids of natural membranes are in the fluid state at the temperatures normally encountered by the organism. The fluid state of bilayers has been extensively studied by several physical approaches. Among the most important of these is electron spin resonance (ESR), a technique in which a phospholipid molecule is "*marked*" by attachment to a chemical group containing an unpaired electron.

In the presence of an applied magnetic field, the unpaired electron in the marker group absorbs energy at certain wavelengths in the microwave range, producing a characteristic absorption spectrum with pronounced peaks. If phospholipids containing the spin marker are introduced into a membrane so that they are initially concentrated in a localised area, collisions among the molecules cause the characteristic absorption peaks to broaden. As the marked molecules diffuse into the surrounding bilayer, collisions between them decrease and the absorption peaks narrow. From the rate at which the spectrum narrows into sharper peaks, the rate of movement or diffusion of the marked molecules through the bilayers can be determined.

From data of this type several groups of investigators have confirmed that phospholipids in gel-phase bilayers are essentially immobile and that the phospholipid molecules of bilayers at temperatures above the phase trrrr ition are free to move. The rate of lateral diffusion determined for individual phospholipid molecules in artificial bilayers in the fluid state is surprisingly high, about 1- 2 $\times$ $10^{-8}$ $cm^2/s$. Measurements of ESR markers added to living membranes give equivalent results, indicating that natural membrane bilayers are in the fluid state, with diffusion rates for individual phosphollpid molecules approaching the same value of $10^{-8}$ $cm^2/s$.

Diffusion rates of this magnitude would mean that a phospholipid molecule in the plasma membrane of a bacterium could move from one end of the cell to the other in about one second! These diffusion rates apply to the *lateral* movement of phosphollpid molecules through the plane of model bilayers and natural membranes, Movement of a molecule from one-half of a bilayer to the other, called *flip-flop,* has different characteristics. Significantly, these characteristics differ widely between artificial bilayers, and natural membranes.

In model bilayers, flip-flop between bilayers halves is highly restricted. Estimated rates for this type of movement are so low that phospholipid molecules can be considered to be restricted essentially to their own half and to diffuse laterally on only one side. Natural membranes, in contrast, show much faster flip-flop rates, probably moderated by membrane proteins specialised for this function. The relative degree of flip-flop is fundamental to the maintenance of lipid *asymmetry* in membranes, a condition in which different proportions of phospholipid types are present in the two halves of a bilayer.

**The Effects of Cholesterol**

Adding cholesterol to the phospholipids used to construct artificial bilayers has a marked effect on the phase transition and membrane fluidity. Addition of small amounts of cholesterol widens the temperature range over which the bilayer melts. Sufficiently large amounts (artificial bilayers will incorporate cholesterol up to a 1 : 1 molar ratio with phospholipids) will eliminate the phase transition altogether.

These effects are belieN·ed to be due to the position taken by *ternl in* hilavers. X-ray diffraction indicates that cholesterol lines up between the phospholipid molecules of a bilayer half, with the long axis of the cholesterol molecule paralleling the hydrocarbon chains of the phospholipids. In this position, cholesterol interferes with the tight packing of hydrocarbon chains necessary for transition to the gel phase. As a result, cholesterol acts as a bilayer "*antifreeze*" and keeps the bilayer fluid at lower temperatures.

Above the phase transition, the presence of cholesterol between the hydrocarbon chains of adjacent phospholipid molecules has the opposite effect and restricts their movement, reducing (but not eliminating) the fluidity of a bilayer at elevated temperatures. These effects are most significant for the plasma membranes of animal cells, which may contain as much as 30% cholesterol by weight, equivalent to one cholesterol molecule for every two phospholipid molecules.

## METHODS FOR ISOLATION

The ideal isolated membrane component is that which does not have any non-membrane contamination. The isolation of the membrane can be regarded as the removal of other cell components from the cell. This operation is somewhat less difficult in the non-nucleate mammalian red blood cells.

During hemolysis the membrane structure loosen sufficiently to allow the cell contents to escape without rupturing completely. After this hemolysis the isolation of erythrocyte plasma membrane is easy. However, the product obtained depends upon the method of lysis and the pH of the liquid.

The other procedure of getting the plasma membrane is introduced by *Wolport* and *O'Neill* in 1962 by treating the *Amoeba* with buffered; 45% glycerol or 2.4 M sucrose solution, caused the cell contents to shrink away from the cell membrane, which retained its characteristic shape.

The membranes were fused from the cell contents by gently homogenisation and were subsequently collected in bulk by centrifuging density.

## MOLECULAR STRUCTURE OF PLASMA MEMBRANE

All biological membranes, including the plasma membrane and the internal membranes of eukaryotic cells, have a common overall structure: they are assemblies of lipid and protein. molecules held together by noncovalent interactions. The lipid molecules are arranged as a continuous double layer 4 to 5 nm thick.

This lipid bilayer provides the basic structure of the membrane and serves as a relatively impermeable barrier to the flow of most water-soluble molecules. The protein molecules are "*dissolved*" in the lipid bilayer and mediate the various functions of the membrane; some serve to transport specific molecules into or out of the cell; others are enzymes that catalyse membrane-associated reactions; and still others serve as structural links between the cell's cytoskeleton and the extracellular matrix, or as receptors for receiving and transducing chemical signals from the cell's environment.

All cell membranes are dynamic, fluid structures : most of their lipid and protein molecules are able to move about rapidly in the plane of the membrane. Membranes are also asymmetrical structures ; the lipid and protein compositions of the two faces differ from one another in ways that reflect the different functions performed at the two surfaces. Although the specific lipid and protein components vary greatly from one type of membrane to another, most of the basic structural and functional concepts discussed in this chapter are applicable to intracellular membranes as well as to plasma membranes.

After considering the structure and organisation of the main constituents of biological membranes-the lipids, proteins, and carbohydrates-we will discuss the mechanisms cells employ to transport small molecules across their plasma membranes and the very different mechanisms they use to transfer macromolecules and larger particles across this membrane.

### The Lipid Bilayer

The first indication that the lipid molecules in biological membranes are organised in a bilayer came from an experiment performed in 1925. Lipids from red blood cell membranes were extracted with acetone and floated on the surface of water. The area they occupied was then decreased by means of a movable barrier until a monomolecular film (a monolayer) was formed. This monolayer occupied a final area about twice the surface area of the original red blood cells. Because the only membrane in a red blood cell is the plasma membrane, the experimenters concluded that the lipid molecules in this membrane must be arranged as a continuous bilayer.

The conclusion was right but it l rned out to be based on two wrong assumptions that fortuitously compensated for each other. On the one hand, the acetone did not extract all of the lipid. On the other, the surface area calculated for the red blood cells was based on dried preparations and was substantially less than the true value seen in wet preparations. Nonetheless, the conclusions drawn from this experiment had a profound influence on cell biology ; as a result, the lipid bilayer became an accepted part of most models of membrane structure, long before its existence was actually established.

**Danielli- Davson model**

*Harvey* and *cole* indicated the existence of protein by studying the surface tension of cells. This led them to propose a lipo-protein model of the cell membrane. According to this model the plasma membrane consists of two layers of lipid molecules as shown in the lipid bilayer model.

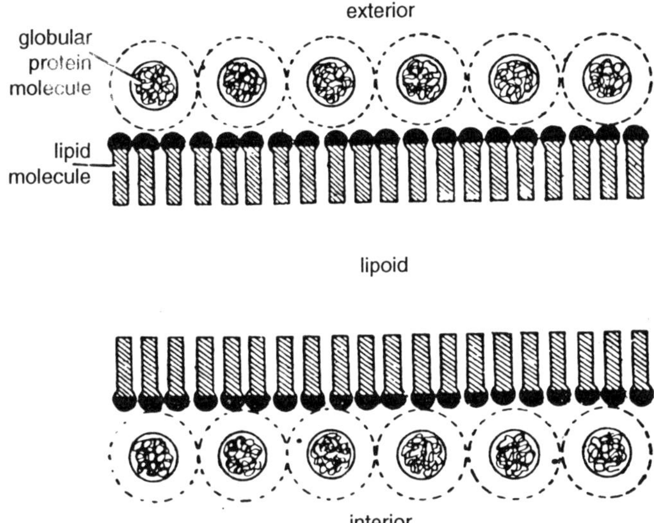


*Figure 3.1: The original Davson-Danielli model.*

The lipid molecules have their polar regions on the outer side.

Globulin proteins are thought to be associated with the polar groups of the lipids. The non-polar hydrophobic ends of the two layers of lipids face each other, whereas their polar hydrophilic ends are associated with protein molecules by electrostatic interation. Protein-linked polar pores are present in the membrane. These pores are formed by periodic continuity of outer and inner layers of proteins of plasma membrane.

Modifications of Danielli-Davson Membrane Model Several adific-ations of the above arrangement have been described

(A) Some plasma membrane have folded α-chains of proteins on both the surfaces of lipid bilayer.

(B) Coiled x-chains of helical protein on the surfaces of lipid bilayer.

(C) With globular proteins on both the surfaces.

(D) With folded proteins on both the surfaces and helical proteins extending into the pores.

(E) With folded α-chain protein on one side and globular protein on the other side.

**Robertson's unit Membrane Model**

Unit membrane model was put forward in the year 1953 pile studying cell under electron microscope. The basic unit iembrane structure was considered to be general for a wide ariety of plant and animal cells. All cell organelles such as rolgi body, mirochondria, endoplamic reticulum, nuclear iembrane etc. have the unit membrane structure. The unit membrane is considered to be trilaminar with a bimolecular lipid layer between two protein layers.

Two parallel outer dense osmiophilic layers of 20A° which correspond to the two protein layers. The middle light coloured osmiophobic layer is about 35A° in thickness corresponding to the hydrocarbon chains of the lipids. Thus the unit membrane is about 7A° in thickness. In this respect it resembles the Danielli-Davson model if, however differs from the Daneilli-Davson model in that the protein in asymmetrical. On the outer surface is mucoprotein, while on the inner surface is non-mucoid protein.

Objections to unit membrane theory, Objections to the unit membrane model increased during the 1960s and this led to reexaminations of lipid-protein interactions and to new models. Studies of *F.S. Sjostrand*, of smooth endoplasmic reticular, mitocbondrial and chloroplat membranes underscored the differences between observed features of membranes and the uniformity that was required by the unit membrane concept.

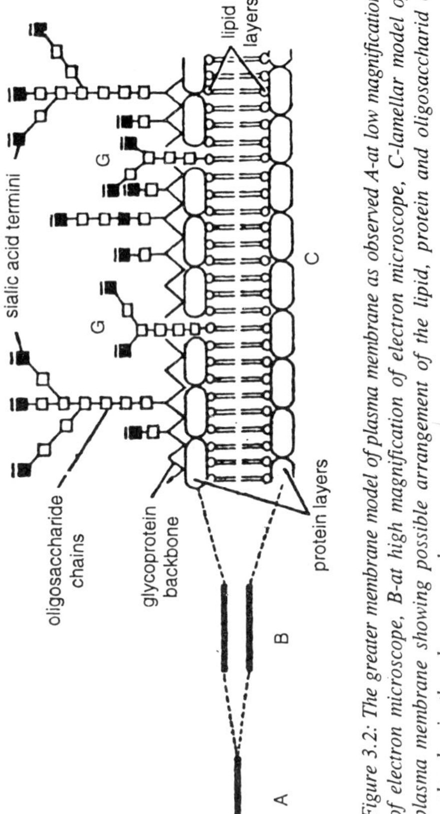


*Figure 3.2: The greater membrane model of plasma membrane as observed A-at low magnification of electron microscope, B-at high magnification of electron microscope, C-lamellar model of plasma membrane showing possible arrangement of the lipid, protein and oligosaccharid e molecules in the plasma membrane.*

Mitochondrial and chloroplast membranes contain displays of particulate units in. or on the membrane. The plasma membrane did not present the same app mitochondrial or chloroplast membranes. It seemed that different models might be needed to described different functional types. This unsuitable approach became unnecessary when a mosaic membrane model was proposed.

***Greater membrane model***

Like the trilaminar model here too the lipid layer is sandwiched between two layers of structural proteins. *Robertson* described the different nature of outer and inner surfaces of the membrane. The inner surface was thought to be covered with unconjugated protein, and the outer surface with glycoprotein, which is superimposed on the structural protein obigosaccharide chains with negatively charged sialic acid terminals are attached to the glycoprotein.

## Micellar Model

An alternative interpretation of molecular structure *of* plasma membrane has been postulated by *Hoffman*. They have suggested that biological membrances may have a non-lamellar pattern, consisting instead of a mosaic of globular subunits known as *micelles,* which have a lipid core and a hydrophilic shell of polar groups. Lipid micelles are possible building-blocks for membranes since they tend towards spontaneous association. In this model of membrane structure the protein components of the membrane may form a monolayer on either side of the plane of lipid micelles.

Individual units of the micellar mosaic might be replaced by individual enzymes molecules or by arrays of enzymes with a precise-three dimensional organisation, enabling specific functions to be 'built in' to the membrane structure. The spaces between the globular micelles are thought to form water-filled pores 0.4 nm (4A°) in diameter, lined partly by the polar groups of the micelles and partly by the polar groups of associate protein molecules.

## Fluid Mosaic Model

This model was proposed by *Singer* and *Nicolson.* According to this concept, the lipid molecules are arranged to form a rather continuous bilayer that forms the structural frame work of plasma membrane. The protein molecules are arranged in two different manners. Some proteins are located exclusively adjacent to the outer and inner surfaces of lipid bilaN er and are called *extrinsic proteins.* Other proteins penetrate lipid bilayer partially or wholly and form *integral* or intrinsic proteins.

The lipids and integral proteins of plasma membrane are *amphipatic* in nature. The term amphipaty was coined by *Hartley, 1936* for those molecules which have both hydrophobic and hydrophilic groups. The amphipatic molecules tend to constitute liquid crystalline aggregates in which hydrophobic or non-polar groups are situated inside the bilayer, and hydrophilic groups are directed towards the water phase.

Therefore, the lipid molecules form a rather continuous bilayer. The integral proteins are interecalated in the lipid bilayer, with their polar regions protruding from the surface and non-polar regions embedded in the lipid bilayer. This arrangement explains why the active sites of enzymes and antigenic glycoproteins are exposed to the outer surface of the membranes. The quasifluid structure of plasma membrane explains the movement of cluster of protein molecules of considerable size across the membrane.

**Pores in Plasma Membrane**

Plasma membrane is perforated by pores. These have a diameter of about 0.35 nm (nanometer), slightly larger than the sodium ions. Less than 0.1 percent of the plasma membrane is perforated by pores while 99.9 percent of the cell surface is impenetrable for ions. Several models of structures of pores have been proposed. Some of them are

1. *Structural pores.* These are permanent cylindrical holes that interrupt the otherwise continuous bilayer sheet.
2. *Dynamic pores.* These pores are transient cylindrical holes rather than being permanant. These appear only at the time of intake.

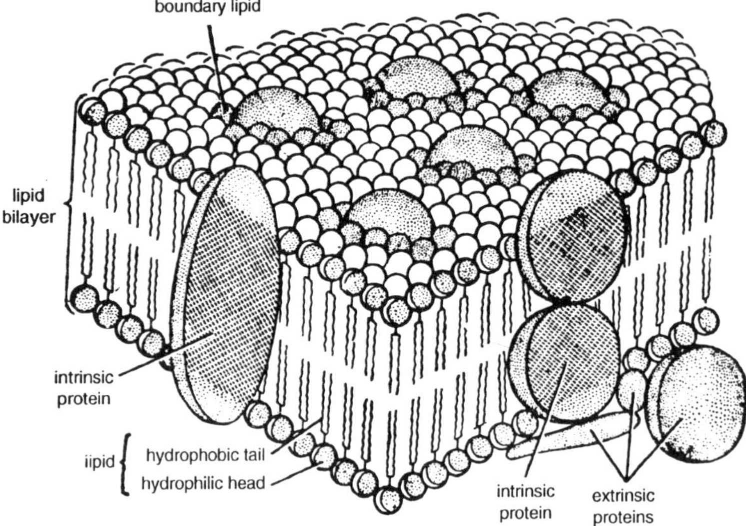


*Figure 3.3: Fluid mosaic model of the plasma membrane.*

3. *Paving block pores.* According to this concept the pores are regarded to be the corners of the closely filled nearly hexagonal paving blocks of lipid and protein subunits.
4. *Protein channel pores.* These pores are considered to be parts of the lipid-globular protein mosaic model. These form small channels of specific proteins embedded in the membrane through which ions and small molecules can diffuse.
5. *Ionophore.* The ionophores are small polypeptides whose one end is hydrophobic and other hydrophilic. The hydrophobic (outer) end dissolves in the membrane while the hydrophilic end (inner side) picks up ions or water soluble materials and

dumps them on the other side. The ionophores help in exchange of substances from or into the cell.

## SPECIALISATION OR MODIFICATIONS

With the increased resolution of the electron microscope numerous specialisation of cell surface have been recognised. Following description of *Fawcett* deals with the various specialisations of plasma membrane studied topographically.

### Microvilli

In the intestinal epithelium microvilli are very prominent and form a compact structure that appears under the light microscope as a striated border. These microvilli, which are 0.6 to 0.8 pm long and 0.1 um in diameter, represent cytoplasmic process covered by the plasma membrane. Within the cytoplasm core fine microfilaments are observed which in the subjacent cytoplasm form a terminal web. The outer surface of the microvilli is covered by a coat of filamentous material (fuzzy coat) composed of glycoprotein macro- molecules.

Microvilli increase the effective surface of absorption. For example, a single cell may have as many as 300 microvilli, and in a square millimeter of intestine there may be 200,000,000. The narrow spaces between the microvilli from a kind of sieve through which substances must pass during absorption. Numerous other cells, in addition to intestinal epithelium, have microvilli, although fewer in number.

They have been found in mesothelial cells, in the epithelial cells of the gallbladder, uteras, and yolk sac, in hepatic cells, and so forth. The brush border of the kidney tubule is similar to the striated border, although it is of larger dimensions. An amorphous substance between the microvilli gives a periodic acidSchiff reaction for polysaccbarides. Between the microvilli, at the base, the cell membrane invaginates into the apical cytoplasm. These invaginations are apparently pathways by which large quantities of flow enter by a process similar to pinocytosis.

### Desmosomes or Macula Adherens

Desmosomes are cell junctions found mainly in the cells of simple columnar epithelium. These occur as specilised areas along the contact surfaces. Under light microscope the desmosomes are seen as darkly stained bodies. Under electron microscope these appear as button-like thickenings on the inner surface of plasma membranes of adjacent cells at the point of contact. The thickenings are traversed by fine cytoplasmic fibrils called *tonofibrils,* which form a kind of loop in a wide arc.

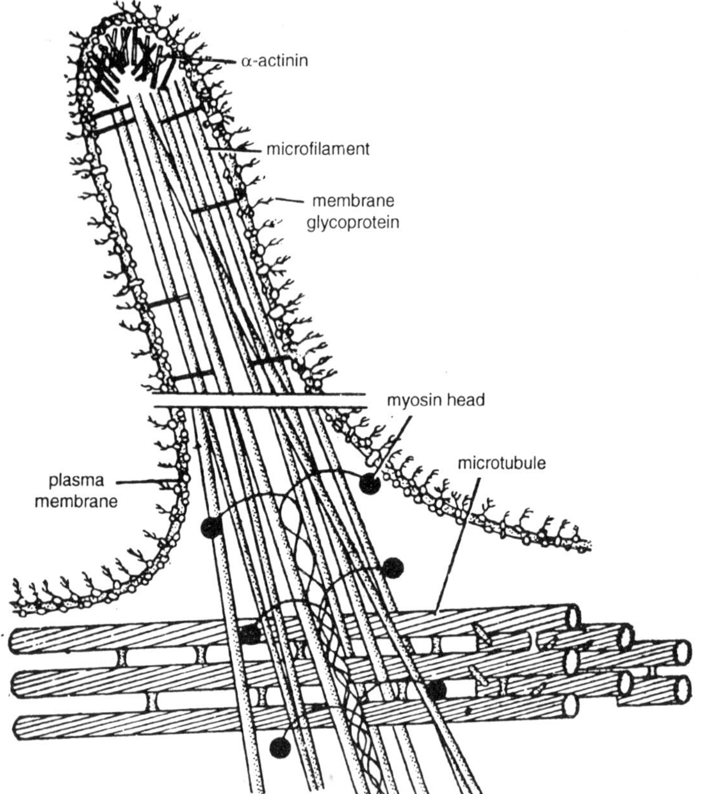


***Figure** 3.4: Ultrastructure of a microvillus showing its different cytoskeleton components. **Note** glycoproteins embedded in the plasma membrane and connections between **microfilaments** and membrane proteins.*

These filaments stabilise the junction and act as anchoring sites for the cytoplasmic structures. The plasma membranes of adjoining cells in the regions of desmosomes are separated by an intercellular space about 30-35 nm. It is filled with intervening dense coating material that forms a dark line in the middle.

It is formed of mucopolysaccharides and proteins. The desmosomes are primarily concerned with cell adhesion, but also help in maintaining cell-shape, providing it rigidity and cellular support. The former is brought about by the intercellular coating substance and latter by the tonofibrils.

**Plasmodesmata**

Sometimes, the cells are joined by bridges of cytoplasm passing between pores of cell wall or plasma membrane between the adjacent cells, such connections are called plasmodesmata. They are usually simple but anastomosing plasmodesmata may also found. Their distribution and number may also very considerably. They were discovered by *Tangl* and were named as such by *Strasburger*. Endoplasmic reticulum often is closely associated with the cell surface, at the points where plasmodesmata are present. Through them cytoplasmic continuity is often maintained among the adjacent cells. They provide a mean for interaction between adjacent cells which are separated in other regions.

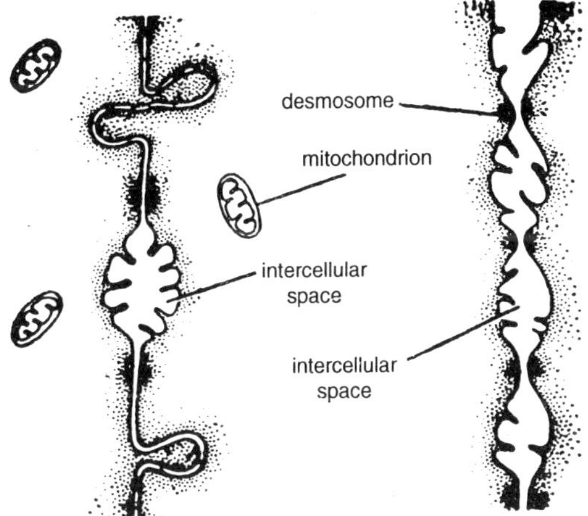


*Figure 3.5: Inter-cellular space.*

Through them the material can pass form cell to cell. It is not known whether all plasmodesmata are similar to each other. There exist some difference because they are not only produced at the time when cell divides but also formed spontaneously between cells that have grown into contacts with one another, e.q. tyloses in xylem vessel elements. They may occur singly or they can be aggregated into groups. In many primary walls, the plasmodesmata are usually associated with a reduced desposition of wall material, and the area then is known as primary pit or field.

**Hemidesmosomes**

These are found in the basal surface of some epithelial cells. Their structure is similar to desmosomes but these are represented by one half ; their counterpart is usually represented by collagen fibrils.

**Terminal bars**

The terminal bars are also known as intermediary junctions or *zonula adhaerens*. The terminal bars are similar to desmosomes except they lack in the tonofibrils. In terminal bar the plasma membrane is thickened and the cytoplasm of thichened area is dense. The terminal bars occur in the intermediary portion of the plasma membrane of columnar cells just below the surface. The correct identity of zonula adhaerens is still questionable.

**Membrane Interactions**

Another aspect of cell membranes that deserve discussion is the interaction between membranes of different cells. Intercellular communication is important in many cell functions and especially during development of the organism, when cells are constantly interacting with other cells. The nature of membrane interactions may vary from complete cytoplasmic bridges between cells to localised areas of membrane junctions that may involve an area of contact as small as a few angstroms or as large as several micrometers.

The structural nature of the actual contact generally falls into one of three categories ; gap junctions, tight junctions, and septate junctions. *Gap junctions* appear as multilayered structures when observed with the electron microscope. They appear to be two unit membranes closely apposed to each other with a 20 to 40 A° gap between. The total thickness of the entire gap junction is 170 to 190 A°, and they are found in both vertebrates and invertebrates.

They are not found in skeletal muscle fibers or red blood cells. *Tight junctions* are found only in vertebrates and occur in cells such as epithelial cells. These junctions appear to be true fusions between the two membranes, and they are 100 to 140 A° thick.

*Septate junctions* have been found only in invertebrates. They are much larger than the other types of junctions and are characterised by electron-dense cross bridges that extend between the two cell membranes.

## PERMEABILITY OF MEMBRANES

The membranes of a cell pass small ions and molecules through them. The passage of ions or molecules may occur as passive diffusion, or active transport evolving the expenditure of energy. In passive diffusion, membranes may be classed according to their degree of permeability.

(1) *Impermeable. A* membrane of this kind allows nothing to pass through it. Certain unfertilised fish eggs, such as trout, are permeable only to gases, water labelled with deuterium does not peneterate the egg.

(2) *Semipermeable. No* cell membranes are in this categoryk model membrane may be constructed to allow passage of eater molecules, but no solute particles.

(3) *Selectively permeable.* Most membranes of the cell belong to this category. Such membranes alow water and certain selected ions and small molecules to pass through, but prohibit other ions as well as small and large molecules.

(4) *Dialysing membranes.* The endothelial cells and their basement membranes of the capillaries and nephron can act as a dialyser. In this way hydrostatic pressure forces water molecules and crystalloids across the membrane down their concentration gradients while restricting the passage of colloids.

## THE FUNCTION OF CELL MEMBRANES IN TRANSPORT

Substances move constantly in both directions across cellular membranes. Metabolites, including all necessary fuel substances and raw materials, enter the cell from the outside, while waste materials and cell secretions travel in the opposite direction. These movements, moderated by the plasma membrane, maintain the concentrations of molecules inside cells at the levels required for life and cellular functions.

Within eukaryotic cells additional membrane systems, such as those surrounding mitochondria, chloroplasts, and the nucleus, maintain the interior of these organelles as separate compartments with their own distinct molecular solutions. The relative rates by which ions and molecules of various kinds pass to and from the cell and its internal compartments reflect the properties of both the lipid and protein components of membranes.

Penetration through the lipid component of membranes is largely passive and proceeds in response to gradients in concentration. This passive transport is influenced by the hydrophobic, nonpolar character of the lipid component of membranes. In general, nonpolar and hydrophobic molecules diffuse through membranes more readily than hydrophilic and polar or charged substances. Transport moderated by the protein component primarily involves movement of hydrophilic, polar, and charged molecules and ions.

In contrast to transport through the lipid part of the membranes, much of this protein-moderated movement is active ; that is, it proceeds against gradients and requires the expenditure of cellular energy. The active and passive transport associated with membrane proteins has the additional characteristic of specificity: Only certain molecules and ions pass to and from cellular compartments via the protein component of membranes.

The active and passive movement of charged particles through membranes of the restriction of these particles to one side of a membrane produces electrical effects in association with the cell surfaces. Some cells, such as the nerve cells of animals or the cells of the electric organs of fish, utilise and moderate these currents or charges as the basis for their specialised functions in communication or defense.

### The Effects of Semipermeable Membranes of Diffusion

An artificial or natural membrane placed between two regions containing particles of different chemical or electrical potential will have no effect on the final outcome if all of the molecules or ions can pass through the membrane with equal ease. However, the net movement may be altered, sometimes in unexpected ways, if the membrane allows some of the molecules or ions to pass through much more readily than others or excludes some molecules or ions entirely. Such a membrane is said to be *semipermeable.*

All biological membranes have this property and thus influence the net movement caused by gradients of chemical and electrical potential. The phenomenon *of osmosis* provides a familiar example of the effects of introducing a semipermeable membrane between two regions containing molecules of differing chemical potential. Consider two compartments of the same volume containing water molecules and separated by a barrier that allows water molecules to pass.

If both spaces are at the same temperature and pressure, and water molecules can pass across the barrier from either side at the same rate, there will be no net movement of water between the sides. Now consider the system, in which the right compartment also contains a quantity of molecules that are larger than the water molecules and cannot pass through the barrier.

Such a system could be duplicated by separating two spaces by a cellophane barrier and placing pure water on one side and a solution of proteins in water on the other. Obviously, if the total number of molecules on either side of the semipermeable barrier is the same, the left side has fewer water molecules per unit volume than the right, because some of the total in the right side is made up of the nondiffusing, larger molecules. Therefore, a concentration gradient can be said to exist for the water molecules in this system. More precisely, since the presence of any dissolved substance lowers the chemical potential of a solvent, the water in the right side is of lower chemical potential.

In response, a net movement of water molecules from left to right will occur, even though the initial volume, pressure, and absolute temperature of the two spaces are the same. Osmosis refers to movement of water molecules in response to a potential gradient of this type. Since osmotic movement of water occurs in response to a gradient of chemical potential, it can accomplish work. This can be demonstrated by the apparatus.

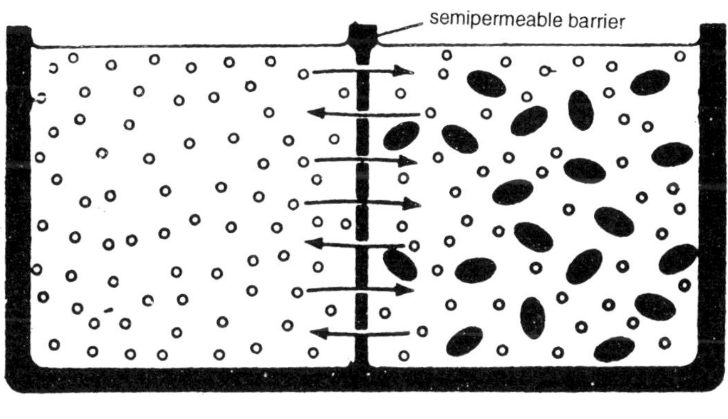


*Figure 3.6: Osmotic flow of water in a system in which a semipermeable barrier separates two compartments.*

The apparatus consists of a tube containing a solution of proteins in water. The bottom of the tube is covered by a sheet of cellophane, sealed tightly to prevent leakage. The tube is suspended in a beaker of pure water. The level of the solution in the tube will rise as water molecules move from the surroundings into the tube in response to the gradient in chemical potential.

When the pressure created by the weight of the raised solution in the tube exactly balances the tendency of water molecules to move from outside to inside, the level will become stationary. At this point, the system i; in equilibrium, and no further net movement of water will occur. The pressure required to leaves of the plant. On the other hand, protista and other small cells or organisms living in fresh water must expend energy to excrete the water constantly entering by osmotic movement to keep from bursting.

Some types of cells, such as those of bacteria, blue-green algae, and almost all plants, are kept from bursting by a thick cell wall. Cells living in surroundings containing highly concentrated salt solutions have the opposite problem and must constantly expend energy to replace the water lost to the outside by osmosis.

### The Effects of Membrane Lipids and Proteins on Passive Transport

The cellophane film used as a semipermeable membrane in i apparatus designed to demonstrate osmosis acts essentially a uniform molecular sieve, with no chemical or physical Fects beyond restricting passage of molecules beyond a certainise. However, the lipids and proteins of biological membranes, and chemical properties, produce a semipermeable barrier that greatly modifies the diffusion of molecules due to potential gradients.

The effects of membrane lipids and proteins an diffusion ·ere first analysed at the turn of the century by E. Overton, who studied the behaviour of substances as they penetrated lant and animal cells. Overton observed that the major ifference between diffusion across biological membranes and iffusion across artificial barriers such as cellophane is that pid solubility modifies the rate of penetration into cells. Generally, the more soluble a substance is in lipids, the more spidly it penetrates into living cells, up to a limit determined y molecular size. Overton's work provided the first clue that ells are surrounded by a surface layer of lipids. These findings were considerably expanded in 1933 by the experiments of R. Collander and H. Barlund, who compared the lipid solubility of more than 30 different substances with the amounts penetrating per unit time across the plasma membrane of cells of *Chara, an* alga. Lipid solubility was evaluated by determining the lipid-water *partition coefficient* for each substance:

$$\text{partition coefficient} = \frac{\text{amount dissolving in a lipid}}{\text{amount dissolving in water}}$$

Olive oil was used as the solvent for determining lipid solubility. By plotting the permeability (amount penetrating per unit area of cell surface per unit time) against the partition coefficient for each substance, Collander and Barlund confirmed that the penetration of substances into cells is directly related to lipid solubility : the more lipid soluble, the more permeable. On this basis, they assumed that penetrating dissolve in the lipid of the plasma membrane in order to pass from outside to inside.

Later work revealed that this conclusion was an over simplification and that the proteins of membranes, unknown to Collander and Barlund,

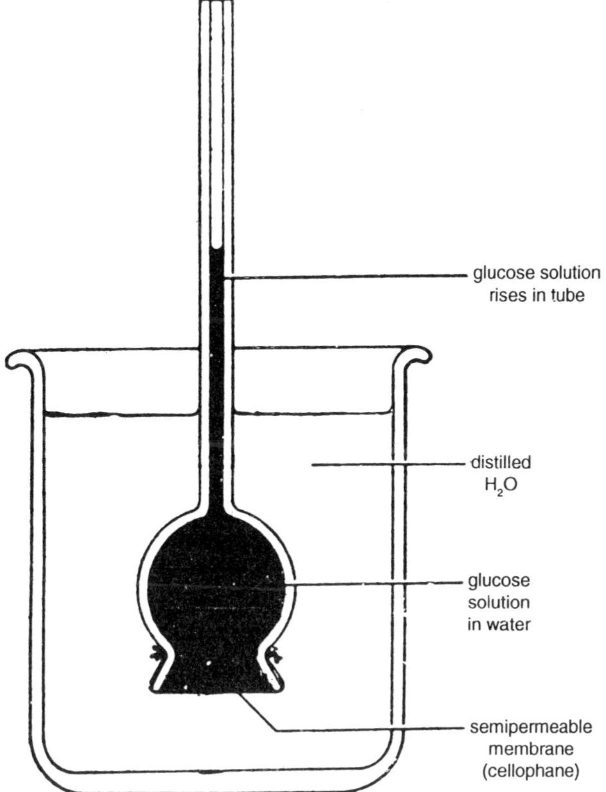


*Figure 3.7: Apparatus demonstrating the ability of osmotic flow to accomplish work. Water will flow osmotically from the surroundings into the tube, raising the level of the solution in the tube. Flow continues until the column of water (d) develops sufficient pressure in the solution to exactly counterbalance the tendency of water molecules to flow inward in response to the gradient in chemical potential.*

also affect permeability. When artificial mem branes consisting of single phospholipid bilayers could be constructed Collander investigators and Barlund penetrate through substances studied by such membranes much more slowly than through natural membranes.

Only the most hydrophobic molecules, with relatively high partition coefficients, penetrate as fast. Few biological molecules other than the long-chain fatty acids and sterols fall into this category. Several important metabolites, including glucose, urea, glycerol, and some of the amino acids, which pass through natural membranes rapidly, penetrate artificial phospholipid bilayers only very slowly or not at all. The differences noted in permeability were greatest for charged particles. $Cl^-$ ions, for example, were found to penetrate artificial lipid films at a rate equivalent to only 1.7 × $10^{-18}$ mol/ cm/s. The permeability of red blood cells to this ion is much higher, equivalent to a rate of 1.4 × $10^{-8}$ mol/cm/s.

These findings indicate that proteins modify natural membranes by increasing the solubility and rate of penetration of hydrophilic substances. Water is an interesting exception to these conclusions because it penetrates both artificial and natural membranes quite rapidly. If assigned an arbitrary low partition coefficient, water penetrates living cells such as *Chara* much faster than expected, falling considerably above the essentially straight-line plot obtained for the other molecules. Penetration of water through phospholipid films takes place at a comparable rate. The basis for this exceptional behaviour is unknown.

**Membrane Proteins and Facilitated Diffusion**

Further investigations into the rapid penetration of substances like glucose, urea, and glycerol has provided insights into the mechanisms by which membrane proteins enhance passive transport. The degree of enhancement is sometimes large. Glucose at low concentrations, for example, penetrates passively into cells about 100,000 more rapidly than expected from its partition coefficient.

Although rapid, this enhanced passive penetration still depends on diffusion. The energy required for transport is provided by a favourable gradient in a chemical potential and does not require an expenditure of cellular energy. Enhanced transport that follows gradients but proceeds at a rate significantly higher than predicted from the partition coefficient is called *facilitated diffusion.*

Measurement of the rate of penetration of substances transported by facilitated diffusion reveals a fundamental' characteristic : At successively higher concentrations, the degree of enhancement drops off until at some point further increases cause no further rise in the rate of penetration. This is in sharp contrast to the behaviour of molecules that penetrate according to their partition coefficients.

For these molecules, permeability increases regularly with concentration, with no dropoff at higher levels. The dropoff noted for facilitated diffusion at high concentrations closely resembles the behaviour of enzymes in catalysing biochemical reactions: As the substrate concentration increases, enzymes gradually become "saturated" and the rate of the reaction levels off.

The similarity in behaviour between the two systems indicates that facilitated diffusion is moderated by carrier molecules in natural membranes with properties similar to those of enzymes. For the same

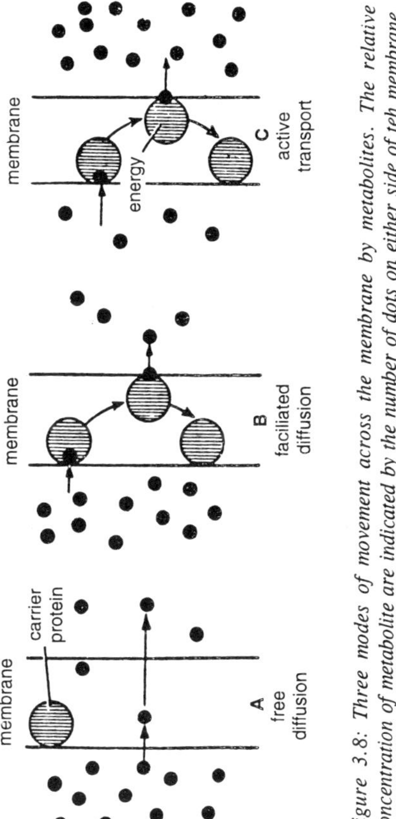


*Figure 3.8: Three modes of movement across the membrane by metabolites. The relative concentration of metabolite are indicated by the number of dots on either side of teh membrane.*

reason, the membrane carriers are assumed to combine briefly with the transported substances as a part of the facilitation mechanism. Other characteristics of the process also point to the involvement of carrier molecules resembling enzymes in facilitated diffusion.

The mechanism exhibits *specificity*. Each molecule transported by facilitated diffusion is carried by a separate mechanism specific only for that substance and closely related molecules. For example, the carrier system facilitating the transport of glucose will also transport the closely related sugars mannose, galactose, xylose, and arabinose. However, it will carry only the naturally occurring D isomers of these sugars and not the L isomers.

The specificity of facilitated diffusion further resembles the specificity of enzymes in that substances with structures closely related to the normally tranaported molecule can inhibit the rate of transport of that

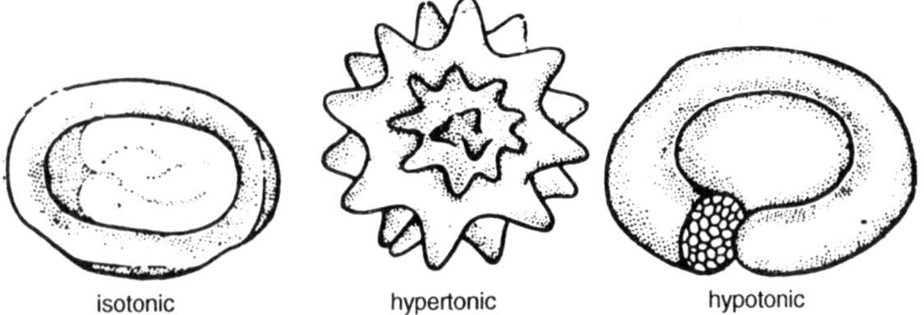


*Figure 3.9: The effects of isotonic, hypertonic and hypotonic solutions on erythrocytes. In isotonic solution erythrocytes maintain their normal shape. Hypertonic solutions cause water to leave erythrocytes, resulted in shrunken cells with a plasma membrane thrown into folds (created). Hypotonic soloutions cause water influx and result in spherical cells that may eventually burst (bemolyes).*

molecule. The similarities observed between facilitated diffusion and enzymatic catalysis indicate that the membrane sites acting on the transported molecules to increase their permeability are proteins with properties similar to those of enzymes.

Of these similarities, the specificity of facilitated diffusion provides the strongest indication that the carrier molecules are proteins, since proteins are the only known molecules that could "recognise" the transported molecules with the required degree of specificity.

**Ionophores and the Carriar Mechanism of Facilitated Diffusion**

Further indications that facilitated diffusion depends on the activity of membrane proteins has come from research with *ionophores,* carried out in the laboratories of H.A. Lardy, B.C. Pressman, P. Mueller, and others. Ionophores are antibiotics that, when added to artificial phospbolipid membranes, greatly increase the permeability of positively charged ions such as $Na^+$ and $K^+$.

Two of these ionophores, *valinomycin* and *gramicidin,* have been most studied. Both resemble protein molecules. Valinomycin consists of alternating amino and hydroxyl acids in a chain; gramicidin is formed from a chain of 15 alternating L and D amino acids. Both molecules take on a cyclic form, with hydrophobic groups exposed on exterior surfaces and oxygcns directed toward the interior.

The interior oxygens line a pore or cavity extending through the inside of the molecule. When wound into its three-dimensional form, neither molecule is large enough to extend completely through the hydrophobic portion of the membrane. When placed in solution with metallic ions, the ionophores form complexes in which the ion is held in the interior of the ionophore through interactions with the oxygens lining the internal cavity.

The ion is then effectively surrounded by the ionophore. The entire outer surface of the complex is hydrophobic in character because of the nonpolar groups directed toward the surface of the ionophore. Supposedly, ,he ionophore forms its complex with the transported ion at the membrane surface, removing it from the surrounding aqueous medium. The ionophore then transports the ion through the hydrophobic membrane interior and releases it to the hydrophilic medium outside the membrane.

Exactly how ionophores accomplish the transfer is unclear, but one or both of two mechanisms are considered possible. In one, the ionophore acts as a mobile carrier, shuttling back and forth between the two membrane surfaces with its enclosed ion.

In the other, several ionophores stack together, forming a continuous pore or channel that extends entirely through the membrane. Experiments indicate that either mechanism may operate, depending on the ionophore. If phospholipid membranes are "frozen" by reducing their temperature below the phase transition, the activity of valinomycin in increasing the permeability of ions is greatly inhibited or eliminated. The activity of gramicidin is unaffected by this treatment. This suggests that valinomycin is a mobile carrier and gramicidin a pore or channel former, because reducing the fluidity of the membrane interior should inhibit movement of a mobile carrier but not affect the structure or function of a pore.

Information from other sources supports these conclusions. Valinomycin can increase movement of ions across an interface between toluene and butanol that measures on the order of micrometers. It is *difficult* to imagine how this process could occur by any other means than shuttling by the ionophore. Other experin-ents support the channel-forming activity of gramicidin. S.B. Hladky and D.A. Haydon followed changes in electrical current between two solutions containing $K^+$ ions exposed to a voltage difference and separated by an artificial membrane containing gramicidin (current is a measure of the number of ions moving from one solution to the other).

The current in this situation was found to fluctuate in jumps, moving rapidly from one value to another in stepwise fashion. The jumps occurred much more rapidly than a molecule the size of gramicidin could move completely from one membrane surface to the other. This suggested a mechanism involving movement of a gramicidin molecule the 1 nm or so required to complete or interrupt a stack rather than diffusion of the iono. phore across the entire width of the membrane.

The proteins carrying out facilitated diffusion in natural membranes may also act by one of the same mechanisms, as either mobile carriers or channel formers. Of the two possibilities, channel formation seems more compatible with the SingerNicolson fluid mosaic model for membrane structure. According to the model, all of the integral membrane proteins have polar and nonpolar surface regions anchoring them in the membrane.

For a membrane protein to act as a mobile carrier, a polar region directed toward one surface would have to submerge in the nonpolar membrane interior to reach the other side. This would be true whether the membrane protein facilitates diffusion by shuttling from one surface to the other or by rotating or tumbling on its axis. In his consideration of the thermodynamic implications of the fluid mosaic model, Singer has calculated that the energy required to submerge polar groups in the nonpolar membrane interior to facilitate diffusion would be prohibitively high, much higher than the amounts available from gradients in chemical or electrical potential. Channel formation, on the other hand, can easily be reconciled with the structure of membrane proteins since most are large enough to extend entirely across the membrane.

The limited experimental evidence available on membrane transport proteins supports these arguments against mobile carriers. There is evidence that the movements in membrane proteins during transport are so small that they probably involve changes in position of restricted segments of the proteins rather than rotation or diffusion of entire molecules. This was demonstrated by recent experiments in Singer's laboratory, in which an antibody molecule was attached to a portion of a transporter protein exposed at the membrane surface.

Attachment of the antibody would be expected to prevent or greatly inhibit any overall rotation or shutting movement of the transporter protein, since any such movement would force the attached antibody protein through the membrane. In spite of the attached antibody, no inhibition of transport was noted, indicating that the protein remains essentially fixed in position in the membrane.

These results effectively rule out rotation or movement of entire transporter proteins between the membrane surfaces. Thus, channel or pore formation remains as the likely mechanism for facilitated diffusion. In this model, the protein, which extends entirely across the membrane, remains fixed in its orientation toward the polar and nonpolar regions of the membrane. The folding of the protein's amino acid chain may create a pore or channel through the interior of the molecule, or two or more protein molecules may align side by side to create a channel between them. During transport, a polar "*active site*" located in the channel and directed toward the membrane surface combines with the transported substance.

Combination causes a conformational change in this limited portion of the protein molecule, causing translation of the active site through the channel to the other side of the membrane. This conformation reduces the affinity of the active site for the transported substance, and it is released. Upon release of the substance, the protein molecule reverts to its original conformation, with the active site exposed in a position to combine with another molecule.

Calculations of the energy required for this movement are within the amounts available from commonly observed gradients in chemical or electrical potential. The mechanisms of passive transport, involving simple and facilitated diffusion, account for the movement of a wide variety of substances to and from cells. By these mechanisms, cells are able to absorb many of the hydrophobic and hydrophilic molecules required for their metabolic reactions and to release waste materials or secreted products to the outside. These transport processes, since they are driven by potential gradients, require no added energy to proceed.

## ACTIVE TRANSPORT

Cells of all types share a common problem in having to choose from among the many molecules and ions present in their environment exactly those substances needed to maintain life. These substances, to the exclusion of others, must be taken up by cells at rates fast enough to maintain growth, reproduction, or specialised function. For example in the kidneys of animals blood is filtered so that virtually all substances below about 30,000 daltons pass into the urine. Some of these. The

$$\text{sugar} + \text{PEP} \rightarrow \text{sugar—P} + \text{pyruvate}$$

standard stage free energy change for this reaction is extremely favourable, about- 40 kl/mole, because of the properties of PEP outlined. The needed PEP can come directly from glycolysis or it can be regenerated from pyruvate by an enzyme unique to bacteria and plants called phosphoenolpyruvate synthetase. The regeneration reaction

$$\text{pyruvate} + P_i + \text{ATP} \rightarrow \text{PEP} + \text{AMP} + PP_i$$

is followed by PP, hydrolysis to pull the equilibrium to the right. This reaction is specifically mentioned here because we will see in the next chapter that it also plays a very important role in photosynthesis.

Group translocation can be broken down into two steps

$$PEP + HP_r \xrightarrow{\text{Enzyme I}} pyruvate + P\text{—}HP_r$$

$$P + HP_r + sugar \xrightarrow{\text{Enzyme II}} sugar\text{—}P + HP_r$$

$HP_r$ is a small, 9500-dalton proteir, a histidine of which gets phosphorylated PEP. The first reaction is catalysed by the cytoplasmic protein, Enzyme I. Transfer of phosphate from HP, to a sugar is catalysed by Enzyme If, which is a complex, membrane-bound protein. Mutant cells lacking $HP_r$ or Enzyme I fail to transport actively any of the sugars handled by this system, but Enzyme II mutations typically affect the uptake of only one sugar. Hence, Enzyme II appears to be a family of proteins, each specific for a different sugar-one for glucose, one for fructose, and so on.

In fact, there is reason to believe that Enzyme II is the carrier protein itself. To understand the mechanism of group translocation or any other transport system, we should know something about the transport proteins involved. Fortunately, some progress in this task has been made, examples of which follow.

**Properties of the Carriers**

To better understand the mechanism of carrier-mediated transport, we should like to be able to examine the responsible carriers in *vitro,* and to use them to reconstruct model transport systems.

***The Isolation of Transport Proteins***

*A* number of molecules that apparently serve as transport proteins have now been isolated, beginning with the galactoside carrier in the mid1960s. One of the first successes was achieved by C.F. Fox and E.P. Kennedy, who took advantage of the fact that, in the absence of galactoside, *E. coli* does not manufacture any significant quantity of the three proteins required specifically and solely for lactose metabolism. When a galactoside .s added to the culture, however, synthesis of the required proteins begins again.

Galactoside permease, the presumed galactoside carrier, is one of the proteins. Its synthesis is inducible- i.e., it is made only in response to need. Inducibility of the galactoside transport system permitted Fox and Kennedy to look for proteins that were present in induced cultures but absent in uninduced cultures. In addition to the other products of the *lac* operon, they were thus able to identify and purify a membrane-associated protein, which they called the M protein, that may be the sought-after carrier.

The M protein was present at about 10,000 copies per cell in the induced state, had a molecular weight of around 30,000 daltons, and in the original cells was apparently exposed to the outside surface as judged by its ability to react with externally applied reagents. Since then, a number of other proteins have been isolated that probably serve as carriers for membrane transport, but it is difficult to prove that they actually have this function in vivo because the appropriate assay (measure of biological activity) does not exist *in vitro.*

One can measure their capacity to bind the substance to be transported, but of course a protein does not have to be a membrane carrier to have a specific affinity for a small molecule. Any enzyme, and a host of other proteins also fit that description. However, some of these isolated proteins have been shown to be necessary for transport, even if it should turn out that they themselves are not carriers. For example, several proteins have been isolated from osmotically shocked bacterial cells.

When bacteria are plasmolysed in sucrose and then quickly diluted with water, their rapid reexpansion causes some protein& to be released and some transport capacity to be lost. That some of the released proteins participate in transport was inferred from the observations that :

(1) mutant cells lacking the capacity to transport a particular substance (cryptic mutants) also fail to release the corresponding binding protein when osmotically shocked ;

(2) in cases where ; transport can be, induced, uninduced cells fail to release the binding protein ;

(3) the binding constants between the free proteins and transportable substances have been shown to be virtually identical with the binding constants for transport of these substances in normal cells and

(4) there are reports that incubation of the shocked cells with solutions of some of the released binding proteins restores at least part of a lost transport capacity.

This group of osmotically released proteins includes individual species capable of binding calcium, sulfate, phosphate various amino acids, galactose, and other substances. They are sometimes called *periplasmic binding proteins* because they appear to reside in the periplasmic space between the cell membrane and the outer membrane that is part of the cell wall in gram-negative bacteria. Their function is apparently not transport as such, but to present substrates to the true carriers. Presumably, the hydrophobic nature of the true carriers prevents

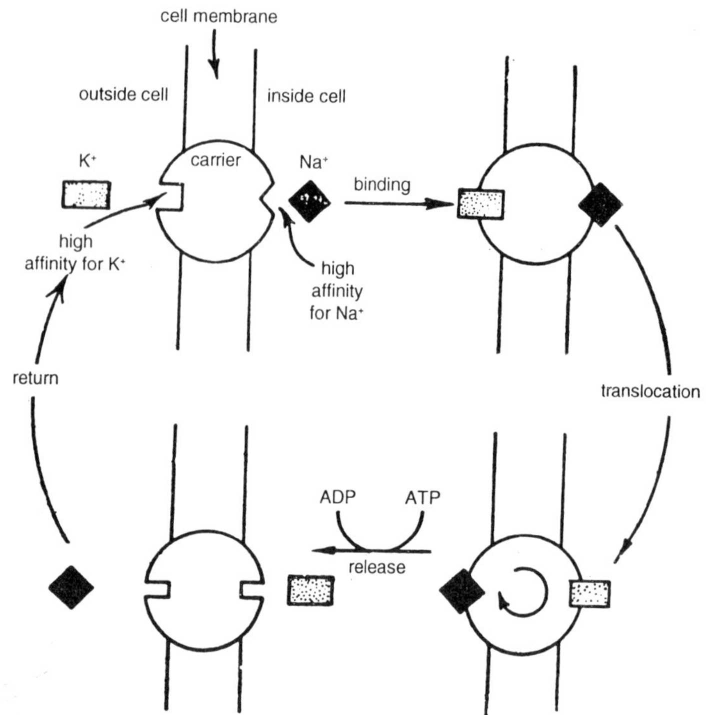


*Figure 3.10: Sodium-potassium exchange pump. Inward transport of $K^+$ ions is accompanied by compulsory outward transport of $Na^+$ ions via carrier proteins within the membrane.*

them from being so easily removed from the cell. Why bacteria have evolved this extra layer of complexity in so many of their transport systems is not certain, but it probably has to do with the difficulty of getting material through their protective but complex cell walls.

There have also been a number of successes in isolating and characterising intrinsic membrane proteins that apparently are true carries. The properties of the $Na^+/K^+$-activated ATPase have already been mentioned. The function of this protein has been proved by incorporating it into artificial membranes and demonstrating ATP-dependent ion translocation. Similar results have been obtained with $Ca^{2+}$- activated ATPase from the endoplasmic reticulum of muscle, with a $H^+$-translocating ATPase from mitochondria, and so on. It is only a matter of time, therefore, before the mechanism of translocation is undertood in more detail.

**The Advantages of Protein Carriers**

From the foregoing observations and arguments, we must conclude that carrier-mediated transport across cell membranes exists and that at least some of the carriers involved are proteins. One might, however, wonder why proteins should be chosen for this role, since they are relatively large and seemingly "*expensive*" to make. There are, of course, very good reasons for that choice

(1) Proteins are specific in their capacity to bind other molecules. They may be given exactly the right shape, with precisely the right distribution of positive and negative charges, of hydrophobic and hydrophilic regions, and of reactive group of various sorts to allow them to make fine distinctions between those molecules that will be bound and those that will not. This specificity can be of great importance to the cell, for a close regulation of metabolic efficiency can only be achieved if the entry and exist of molecules are effectively controlled. Thus, it is possible for the cell to provide a mechanism for a particular molecule to get in and out while in the presence of much larger quantities of other kinds of molecules. Furthermore, the rate of transport is easily controlled by regulating the number of specific carriers

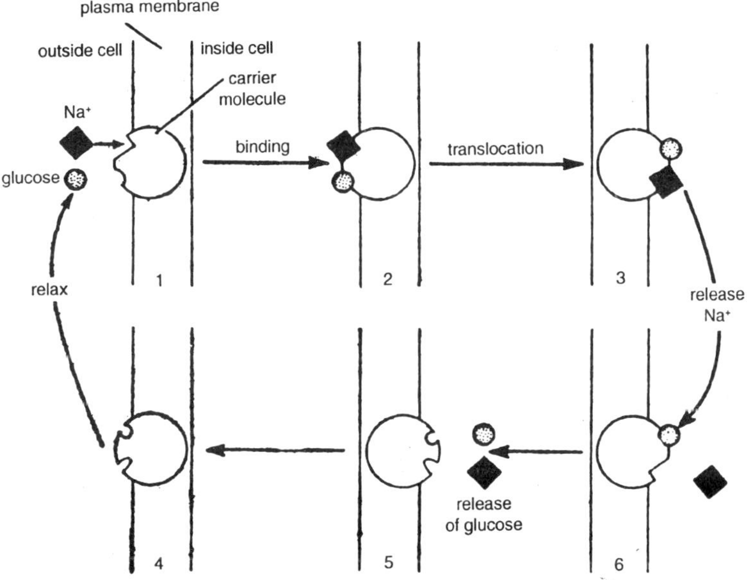


*Figure 3.11: Electrogenic sodium pump. Diagram showing the action of an electrogenic sodium pump through which sugars and other metabolites enter the cell along an electrical gradient generated by the energy-requiring extrusion of $Na^+$ ion.*

irrespective of the rate of transport of other molecules gaining entry by their own mechanisms.

(2) The capacity of proteins to alter their binding affinity for substrate has some important applications when this function is present in a transport molecule. In $Na^+$-coupled transport, for instance, we have a requirement for cooperativity ; specifically, affinity of the carrier for glucose at its "*active*" site is dependent on the presence of sodium ion at its "*regulatory*" site.

(3) Proteins, as catalysts, also make active transport possie by allowing an obligate coupling between two reactionstergetically unfavourable transport with an energetically favourable reaction such as the hydrolysis of ATP. The sodium pump, for example, depends on this kind of coupling.

And finally, it should be pointed out that the use of proteins for transport is actually a conservative choice, for they are the direct products of genes. When small organic molecules are chosen as carriers, their synthesis is apt to require not just one, but several proteins in the form of enzymes. That is, the production of a protein carrier may require only one gene, whereas establishing a biosynthetic pathway to produce any other kind of carrier might require several enzymes, and therefore several genes.

**Nonprotein Carriers**

The suitability and economy of using proleins as carriers does not preclude the use of other organic molecules, for we may easily imagine transport tasks for which proteins are no ideal. Only a few small molecular weight carriers have been isolated. This is not surprising, as they would be present in onl y tiny quantities in the cell and would therefore be difficult to identify. Much of the work on small organic transport molecules has centered on the ionophorous (non-carrying) antibiotics.

These are generally macrocyclic (ringlike) compounds, produced by microorganisms and capable of sequestering inorganic ions. By virtue of its lipid solubility, the ion-antibiotic complex can diffuse through a lipid barrier that would be quite impermeable to the ion itself. One of the more widely studied ionophorous antibiotics is *valinomycin*. It is a 36-atom ring of alternating hydroxy and amino acids, consisting of the sequence [D-valine-D-hydroxyisovalerate-L-valine-L-lactate] repeated three times.

The ionophorous antibiotics are capable of increasing the permeability

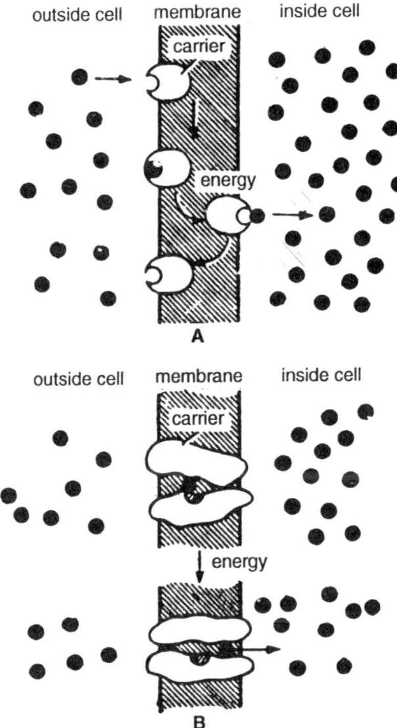


*Figure 3.12: A—The carrier mechanism concept for translocation of hydrophilic molecules across the membrane. B—The conformational change of carrier protein according to the fixed-pore mechanism for translocation of hydrophilic molecules across the membrane.*

of both natural and artificial lipid membranes to small inorganic ions. They often exhibit remarkable selectivity in this role. For instance, valinomycin has about a 10,000 to 1 preference for $K^+$ over $Na^+$, an observation that is not easily explained on the basis of ionic size.

The biological action of the ionophorous antibiotics has been studied with model systems such as simple chloroform barriers in which ordinary salts (e.g., KCl) are quite insoluble. In one such experiment, chloroform was placed in the bottom of a U-tube with a KCl solution above it in one arm and water above it in the other.

Valinomycin, which is capable of transporting a cation-anion pair, was added to the chloroform. Although potassium is the cation favoured by valinomycin, $K^+Cl$ is not a suitable ion pair for the antibiotic. However, picrate, PC or 2, 4, 6-trinitrophenol, was found to be transportable with patassium. Thus, when potassium picrate was added to KCl solution on one side, $K^+Pc^-$ ion pairs were transported across the chloroform by diffusion and concentrated on the other side until a Donnan equilibrium was reached.

$$[K^+]^l\,[Pc^-]^l = [K^+]^r\,[Pc^-]^r$$

Here $Pc^-$ is the picrate anion, and L and R stand for the left and right arms of the U-tube, respectively. This equilibrium relationship was obeyed regardless of the starting concentrations of potassium picrate and potassium chloride on the two sides.

The $Cl^-$ is not picked up by valinomycin and is thereby effectively ignored when the equilibrium is established. Because valinomycin itself is uncharged, it must move both a cation and an anion in order to maintain neutrality. (Charged species, of course, are generally less soluble in lipids.) Other ionophorous antibiotics are charged and may therefore transport a single ion, promoting an ion exchange (e.g., $Na^+$ for $K^+$, $H^+$ for $K^+$, etc.) or creating a voltage gradient, in which case the equilibrium position will follow the Nernst equation.

The principle is the same, however : The antibiotic alone, or the complex of antibiotic and ion (s), is soluble and can therefore diffuse across the lipid barrier. An ion or ion pair is picked up on one side, according to the equilibrium constant between it and the antibiotic, and dropped on the other side according to the equilibrium prevailing there.

Most of the ionophorous antibiotics are products secreted by microorganisms to protect themselves against order microorganisms. They kill by disrupting normal, essential ionic gradients. There are also a few nonprotein carriers that may be secreted but are designed for transport within the cell from which they arise. The most closely studied is a family of bacterial iron-binding substances known collectively as *siderochromes* or *siderophores*.

They are typically secreted from the cell, bind iron, and in that state diffuse back across the bacterial membrane. This may seem inefficient, since obviously many of the carrier molecules will be lost, but in fact bacteria have an absolute requirement for iron, which is typically scarce and for which there is competition from the host. In vertebrates, for instance, iron in the blood is transported tightly bound to a protein called transferrin and therefore is not readily available.

Thus, the bacteria are fighting for survival and produce siderochromes at a rate proportional to their need.

**Membrane Biogenesis**

All indications are that membranes are rapidly assembled and disassembled in cells. Measurements show that the halflife of membrane lipids and proteins is a matter of 15 to about W hours for phospholipids and 50 to 100 or so hours for proteins. Observations of the behaviour of living cells also indicate that membranes are dynamic structures that can be rapidly assembled and disassembled. For example, one investigator has calculated that in the growing cells of a fungus, new membrane is added at the rate of 32 $\mu m^2$ per cell per minute. In spherical cells doubling in size during inter phase growth, new plasma membrane surface area increases approximately 1.6 times.

Where and how are new membranes synthesized ? Answering this question, an old one in cell biology, has been complicated by revelations about membranes arising from the fluid mosaic model and its supporting research. In particular the fact that membrane proteins have strongly hydrophobic and hydrophilic regions raises questions about how and where these proteins are synthesized and how they get into membranes.

How do the hydrophilic portions of proteins that have ends exposed on both sides of a membrane pass through the hydrophobic interior ? How do proteins move from their point of insertion in membranes to more distant locations in the cell ? Essentially the same questions apply to membrane lipids. Where are they synthesized, and how are they transported within membranes ? Recent work has supplied at least partial answers to these questions.

**Evidence on the Origins of Membrane Lipids**

Using radioactively labeled precursors of membrane lipids, D.J. Morre showed that in both plant and animal cells, the first uptake and incorporation of label occurs within minutes in the endoplasmic reticulum (ER). Label then appears in the Golgi complex and lastly in the plasma membrane. These observations suggest that membrane lipids are synthesized in the ER and are first incorporated into membrane bilayers in this location. The newly synthesized lipids then move through the Golgi complex to the plasma membrane.

This conclusion is supported by electron microscopy, which shows that membrane segments "flow" from the ER to the Golgi complex and then to the plasma membrane through the medium of small vesicles that pinch off from one membrane type to fuse with the next. Flow from the rough ER to the nuclear envelope is also possible, since connections between these membrane systems are also frequently seen. The fluid nature of the membrane bilaver probably provides the basis for this lipid flow through the cytoplasmic membrane system.

Localisation of the enzymes required for membrane lipid biosynthesis also supports the conclusion that initial synthesis and incorporation take place in the ER. Much of this work has been accomplished by breaking cells open and separating membranes into fractions by centrifugation. In membranes isolated in this way, the enzymes required for the terminal steps of phospholipid synthesis prove to be associated with the ER fraction.

Enzymes catalysing the final steps of cholesterol synthesis can also be identified in the ER. Membrane flow along the ER-Golgi complex—plasma membrane or ER-nuclear envelope routes readily explains how lipids, newly synthesized in the ER, are transported between these membranes. However, it does not explain how the membranes of mitochondria and chloroplasts, which rarely show connections of any kind to other cellular membranes, receive their lipids. These organelles contain enzymes only for synthesis of a few minor membrane lipids ; most of their membrane lipids apparently originate from the surrounding cytoplasm.

A series of experiments by K.W.A. Wirtz and D.B. Zilbersmit and others has shown that lipid flow between the ER and totally disconnected organelles such as mitochondria is probably promoted by soluble cytoplasmic proteins called e,.change proteins. Wirtz and Zilbersmit found that mitochondria took up label very rapidly when cells were exposed to labeled lipid precursors, indicating that the lipids newly synthesized in the ER are rapidly transferred to mitochondria even though direct membranous connections cannot be detected between the two membrane systems. Subsequently, Wirtz and Zilbersmit isolated a group of soluble cytoplasmic proteins, the exchange proteins, that can promote transfer of lipid molecules between isolated ER and mitochondria and, within mitochondria. between the inner and outer mitochondrial membranes.

Others have also demonstrated exchange proteins that stiulate movement of lipids between the ER and the plasma embrane. These proteins evidently provide the means for )id transfer between the ER, mitochondria, and chloroplasts id also a mechanism for lipid transfer between the ER, Golgi mplex, plasma membrane, and nuclear envelope systems.

The work accomplished to date indicates that the exchange proteins can transfer lipid molecules only to the bilayer half ey face. Subsequent transfer to the opposite bilayer half by flip-flop is probably moderated, as noted earlier, by proteins rming an integral part of the membrane.

**Origins of Membrane Proteins**

Experiments tracing the incorporation of labeled amino idc havo also implicated the ER in the synthesis of membrane oteins. Franke showed that within minutes of exposure to dioactive amino acids in animal cells, label could be detected membrane proteins of the rough ER. Label subsequently )peared in the Golgi complex and, after about 20 min, in the asma membrane. Equivalent experiments by Morre with ant tissues showed the same pattern, with first incorporation 'label in the rough ER, followed by label of the Golgi comex and plasma membranes within 30 to 60 min.

Exposing cells to precursors of carbohydrate groups, such s labeled glucose, shows that the addition of complex sugars o proteins to form glycoproteins occurs in both the ER and he Golgi complex, with greatest concentration of activity in he Golgi complex. Analysis of enzymatic activity in isolated membrane fractions also shows that the greatest concentration of enzymes attaching carbohydrate groups to proteins occurs in the Golgi complex.

These results indicate that proteins, like membrane lipids, are first incorporated into cellular membranes in the rough ER. The new membrane proteins then flow through the smooth ER and the Golgi complex to reach the plasma membrane. Presumable flow from the rough ER to the nuclear envelope also occurs.

This evidence leaves unanswered questions about bow proteins with hydrophilic and hydrophobic regions are inserted into membranes at their sites of synthesis on ER ribosomes. The answers have been supplied by G. Blobel and his associates as a part of a model Blobel calls the *signal hypothesis.* hiobel and his associates have shown that the first series of amino acids assembled during membrane protein synthesis is primarily hydrophobic.

This produces a hydrophobic segment at the "beginning" end of the new protein. According to Blobel's hypothesis, this segment, the *signal,* causes attachment of the ribosome synthesizing the protein to ER membranes. Once attached, the signal sequence, by virtue of its hydrophobic character, penetrates into the membrane. The remaining segments of the protein follow the signal into the membrane as they are assembled.

Once synthesis is complete, the final folding conformation of the new membrane protein determines its asymmetric location in the membrane bilayer through hydrophobic and hydrophilic interactions with the surrounding membrane lipids. Work by others has shown that the initial portions of the complex carbohydrate groups of membrane glycoproteins are added onto the segments of newly synthesized proteins facing the interior of the ER sacs on the side of the ER membranes opposite the ribosomes. The carbohydrate groups are completed as the membrane segments containing the nascent glycoproteins reach the Golgi complex. The fir wbed carbohydrate units face the inside of vesicles pinching off front the Golgi. Fusion of the vesicles with the plasma membrane places the carbohydrates on the exterior surface of the cell.

About 90% of the proteins of mitochondrial and chloroplast membranes originates from the surrounding cytoplasm. There is some evidence that at least some of these proteins enter the organelles by a process analogous to the signal mechanism. In this case, the signal recognises and binds specifically to the organelles, causing insertion of the newly synthesized proteins into the organelle membranes. The experiments outlined in this chapter indicate that lipids and proteins are structured in membranes in the pattern proposed by the fluid mosaic model. Lipids are arranged in a fluid bilayer ; proteins are suspended as globular or particulate units within the bilayer.

The proteins may be completely free to move through the fluid bilayer as individual units, or they may be fixed in position by interactions among themselves or with elements associated with the membrane such as microtubules and microfilaments. The distribution of both membrane lipids and proteins is asymmetric.

The various types of lipid molecules occur in characteristic and different proportions in the two bilayer halves of the various cellular membranes ; all proteins take up specific orientations with respect to the two membrane surfaces. Both the lipid and protein components of eukaryotic cellular membranes originate in the ER and distribute by membrane flow from the ER to other regions. Lipids can also be distributed to membranes by exchange proteins.

# 4

# MITOCHONDRIA

Mitochondria are found in all eukaryotic cells capable of utilising oxygen. They are thus found in aerobically growing yeast, in protozoa, and in virtually every cell of higher plants and animals. These important organelles have been intensively studied for over a century, yet only in the past few years have we begun to understand how they actually function. And, of course, not all questions have been answered even now.

## DISCOVERY

Perhaps the first description of mitochondria was by A. Kolliker, who in 1857 described them as the "sarcosomes" of muscle. Much later be was able to show that the mitochondria of muscle are individual entities and not connected directly to other parts of the cell. However, only after the application of appropriate staining procedures by Altmann in the latter years of the nineteenth century did it become possible to make detailed descriptions of mitochondrial distribution.

From such studies, Altmann concluded that mitochondria are autonomous organisms living within the cytoplasm of a host cell. Many doubted this theory and held to the modern notion that they are integral parts of the cell itself. Nevertheless, there is now the suspicion that mitochondria are, in fact, the highly :evolved descendants of bacteria that once lived symbiotically in higher cells.

The function of these organelles was much debated during the early decades of this century. In 1912, B.F. Kingsbury proposed that they might be the sites of cellular resplrationthat is, the sites of oxygen

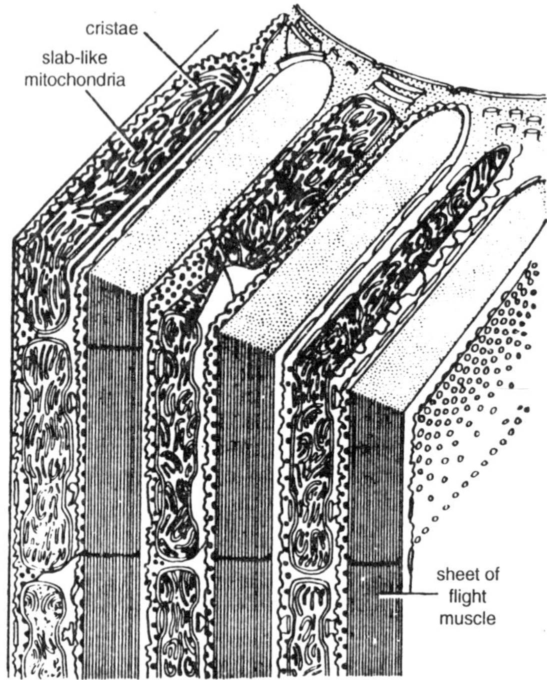


*Figure 4.1: Mitochondrial of the flight muscle of a dragon fly, showing profuse cristae.*

utilisation. He was quite right, of course, though direct proof had to await the development, primarily by George Palade and his associates at Rockefeller University, of improved methods of isolating cellular components. By 1950, however, it was recognised that mitochondria are not only the sites of cellular respiration, but also the major source of ATP production in aerobic animal cells.

**Table 4.1: Some Early Milestones in Respiratory Research.**

| | |
|---|---|
| *ca* 1500 | Leonardo de Vinci likens animal nutrition to the burning of a candle. |
| 1770-74 | Priestley shows that oxygen is consumed by animals. |
| 1780 | Lavoisier and Laplace conclude that the respiration of animals is an oxidation ("burning"). |
| 1857 | Kolliker discovers mitochondria in muscle. |
| 1872 | Pluger finds that oxygen absorbed by the lungs is actually consumed in part by all tissues. |
| 1886 | MacMunn discovers the cytochromes (originally |

| | |
|---|---|
| | "histohematins" ). |
| 1888 | Mitochondria are isolated by Kolliker, |
| 1890 | Altmann'finds a specific stain for mitochondria and suggests that they are autonomous organelles. |
| 1912 | Warburg demonstrates that iron is essential to respiration. |
| 1923 | Keilin shows that cytochromes have an altered oxidation state during respiratory activity. |
| 1928-33 | Warburg determines the basic structure of heme. |
| 1931 | Engelhardt shows that phosphorylation is coupled to oxygen consumption. |
| 1933 | Keilin succeeds in the partial reconstitution of an electron transport chain. |
| 1937 | Krebs formulates the citric acid cycle. |
| 1937-41 | Kalckar and Belitser each devise ways of quantitatively studying oxidative phosphorylation. |
| 1939-41 | Lipmann proposes a central metabolic role for ATP. 1947-50 Lipmann and Kaplan determine the structure of coenzyme A. 1948-50 Kennedy and Lehninger show that the citric acid cycle, oxidative phosphorylation, and fatty acid ox dation to e place in the mitochondria. |
| 1951 | Lehninger shows that oxidative phosphorylation requires electron transport. |
| 1954 | Palade describes the *cristae mitochondriales,* or crista membranes. |

## MITOCHONDRIA

### Distribution

Ordinarily mitochondria are evenly distributed in the cytoplasm. They may, however, be localised in certain regions. In proximal convoluted tubules of the kidney they are found in the basal region of the cell, opposite the renal capillaries. In skeletal muscles they lie between the myofibrils.

In insect flight muscle several large mitochondria are in contact with each fibril. In cardiac muscle the mitochondria are situated in clefts between the myofibrils, numerous lipid droplets are associated with the mitochondria. In many sperms the mitochondria fuse into one or two structures which lie in the middle piece of the tail, surrounding the aril filament. In columnar or prismatic cells they are oriented parallel to the long axis to the cell. In leucocytes they are radially arranged.

**Arrangement of Mitochondria Within the Cell**

Mitochondria often occur in close association with structures that either utilise the ATP they produce or provide the mitochondria with oxidisable substrates. Two striking exa nples of such arrangements are presented. In one, muscle cell mitochondria are seen lined up adjacent to the fibrils that utilise ArP during muscle contraction.

In the other, a mitochondrion is pictured surrounding a lipid droplet that contains fatty acids destined for mitochondrial oxidation. In thin-section electron micrographs mitochondria typically ppear as ellipsoid or oval profiles measuring several micropeters in length and 0.5-1.0 $\mu$m across. The widespread occurrence of the ellipsoid profile has led to the belief that lost mitochondria are sausage shaped, and that typical cells ontain from several hundred to a few thousand such structures. lowever, this commonly held view has been challenged by fans-Peter

Hoffman and Charlotte Avers, whose work on east cells has revealed that the three-dimensional shape of iitochondria within intact cells cannot be inferred from the xamination of a few individual thin-section micrographs. Lfter examining a consecutive series of micrographs obtained y thin-sectioning an entire yeast cell from one end to the ,they, these investigators were able to construct a three:imensional model of yeast mitochondria by assembling all the aitochondrial profiles observed in the individual sections.

The ather surprising result was that the ellipsoid profiles observed n individual thin sections were all found to be derived from ;ross sections through a single large, extensively branched mitochondrion. When this experimental approach was applied to other eukaryotic cells, it was again found that many mitochondrial profiles seen in thin-section micrographs represent portions of larger, interconnected mitochondrial systems. Such results suggest that the number of mitochondria per cell is considerably smaller than once believed.

The conclusion that mitochondria form large interconnected systems is furtfer supported by phase contrast microscopy of living cells, where the problem of sectioning is aliminated. Because the interconnected segments of these large, branched mitochondria appear to be in a state of flux, continually pinching off and re-fusing with one another, the concept of the "*number*" of mitochondria per cell may in fact be meaningless.

In isolated subcellular fractions, mitochondria appear a small separate structures of relatively uniform size. It seem likely, however, that such individual "mitochondria" are arti facts caused by disruption of the branched, interconnecte· mitochondrial network during homogenisation. This type o membrane rearrangement is similar to the fragmentatio process that causes endoplasmic reticulum membranes to b broken up into microsomal vesicles during subcellula fractionation.

## PLASTICITY OF MITOCHONDRIA

*Lewis* and *Lewis* concluded that the mitochondria are extremely variable bodies, which are continually moving and changing shape in cytoplasm. There are no definite types of mitochondria, as any one type may change into another.

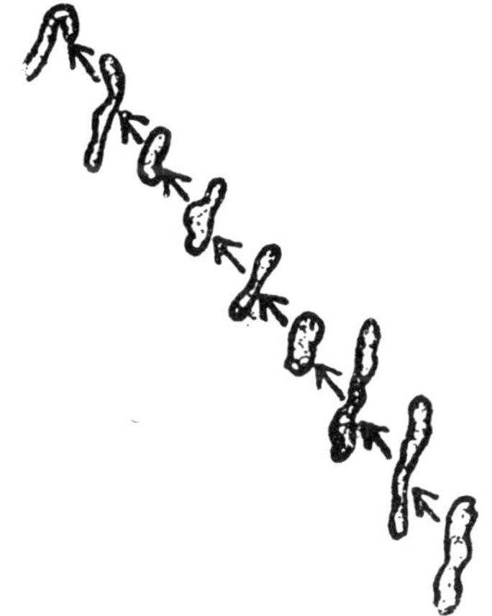

*Figure 4.2: Rapid changes of shape in living cells.*

They appear to arise in cytoplasm and to be used up by cellular activity. The shape may change fifteen to twently times in ten minutes ; it can be changed by heat, hypetonic and hypotonic media or- by acids, fat solvents, potassium permanganate and osmotic changes. *Frederic*, *Littre*, *Tobioka* and *Biesels* studied the effect of large number of chemical and physical agents on mitochondrial behaviour. Some material, such as detergents, show some effect *in vivo* as on mitochondria isolated from homogenates.

## MORPHOLOGY

### Shape

The shape is variable but is characteristic for a cell or tissue type, this too is dependable upon environment or physiological conditions. In general they are *filamentous* or *granular*. They may swell out at one end to become *club-shaped* or hollow out at one end to assume a shape *of tents-racket*. They may become *vesicular* by the appearance of central clear zone. *Rod-shaped* mitochondria are also observable.

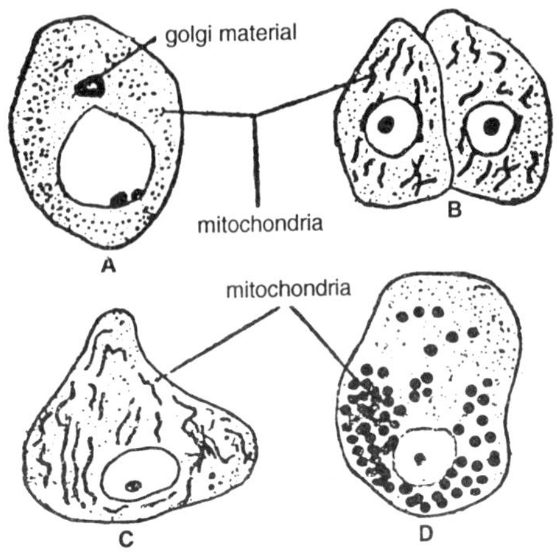


*Figure 4.3: Different shapes and sizes of mitochondria.*

**Size**

The size of mitochondria also varies. In majority of the cells, width is relatively constant, about 0.5p, but the length varies and, sometimes, reaches a maximum of 7$\mu$. The size of cell also depends on the functional stage of the cell. Very thin mitochondria, about 0.2,u, or thick rods 2$\mu$ are also seen.

The size and shape of the fixed mitochondria are determined by the osmotic pressure and pH of the fixative. In acid, pH mitochondria are fragmented and become vesicular. Mitochondria, in the rat-liver, are usually 3.3$\mu$ in length ; in mammalian exocrine pancreas, they are about 10$\mu$ in length and in oocytes of Amphibia, they are approximately of the length 20 to 40$\mu$.

**Number**

Mitochondria are found in the cytoplasm of all aerobically respiring cells with the exception of bacteria in which the respiratory enzymes are located in the plasma membrane. The mitochondria content of a cell is difficult to determine, but in general, it varies with the cell type and functional stage. It is estimated that in liver mitochondria constitute 30 to 35 per cent of the total protein content of the cell and it kidney, 20 per cent. In lymphoid tissue the value is much lower. In mouse liver homogenates there are about $8.7 \times 10^{10}$ mitochondria per gram of fresh tissue. A normal liver cell contains about 1000 to 1600 mitochondria, but this number diminishes during regeneration and also in cancerous tissue.

This last observation may be related to decreased oxidation that accompanies the increase to anaerobic glycolysis in cancer. Another interesting finding is that there is an increase in the number of mitochondria in the muscle after repeated administration of the thyroid hormone, *thyroxin.* An increased number of mitochondria has also been found in human hyperthyroidism. Thus cells with high metabolic activity have a high number of mitochondria, while those with low metabolic activity have *a lower number.*

Large sea urchin *eggs* have 13,000-14,000, while renal tubules have 300-400. In the sperm there are as few as 20-24 mitochondria while in some oocytes there are about 300,000. In the protozoon *Chaos chaos* there are about 500,000 mitochondria. Some algal cells contain only one mitochondrion.

## STRUCTURE OF MITOCHONDRIA

A typical mitochondrion is sausage shaped with an average diameter of about 0.5$\mu$. When it is properly fixed in osmimum containing fluid and studied under electron microscope which reveals that there is hardly any difference between plant and animal mitochondria. In both the cases the mitochondrion is bounded by two membranes, the *outer membrane* and the *inner membrane.*

The space between the two membranes is called the *outer chamber* or *inner-membrane space.* It is filled with a watery fluid, and is 40-70Å in width. The space bounded by the inner chamber is called the *inner chamber* or *inner membrane space.* The inner membrane space is filled with *a matrix* which contains dense granules (300-500Å), ribosomes and mitochondria) DNA. The granules consist of insoluble inorganic salts and are believed to be the binding sites of divalent ions like $Mg^{++}$ and $Ca^{++}$.

In some cases they apparently contain polymers of sugars. The side of the inner membrane facing the matrix side is called the *M-side,* while the side facing the outer chamber is called the *C-side. Two* to six circular DNA molecules have been identified within mitochondria. These rings may either be in the open or in the twisted configuration. They may be present free in the matrix or may be attached to the membrane. The enzymes of the Krebs cycle are located in the matrix.

The inner membrane is thrown up into a series of folds, called *cristae mitochondriales,* which project into the inner chamber. The

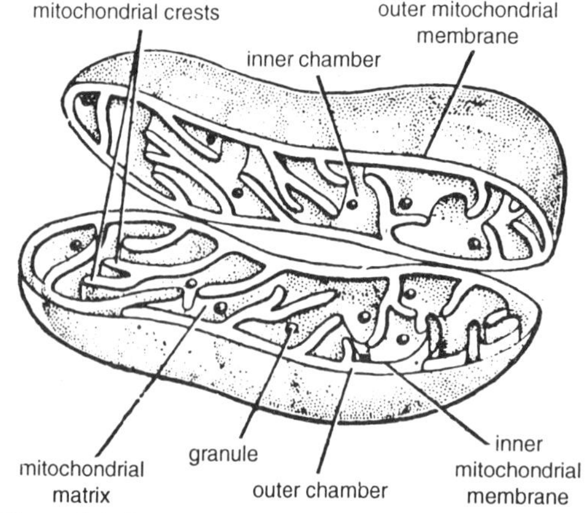


*Figure 4.4: Detailed structure of a typical mitochondrion.*

cavity of the cristae is called the *intercristae space,* and is continuous with the intermembrane space.

## MITOCHONDRIAL CRISTAE

The space and arrangement of crests is variable and may be of the following types

(*i*) Parallel to the long axis of mitochondria as in the neurons and the striated muscle cells.

(*ii*) Concentrically arranged as in the matrix of certain spermatids.

(*iii*) Interlaced to form villi as in *Amoeba.*

(*iv*) Cristae in the form of vesicles *which* form a network of interconnected chambers as in the cells of parathyroid gland and W.B.C. of man.

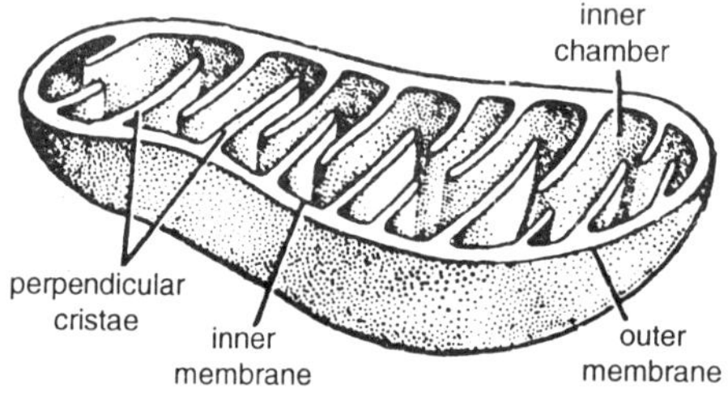


*Figure 4.5: Mitrochondrion with perpendicular cristae.*

(*v*) Arranged in a tubular fashion but perpendicular to mitocbondrial axis as in the cells of adrenal gland.

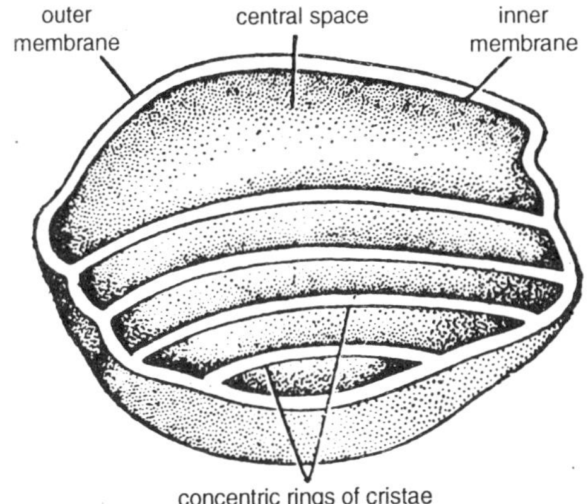


*Figure 4.6: Mitochondrion showing parallel cristae.*

(*vi*) Haphazardly distributed as in the cells of kidney of insects and hepatic cells.

(*vii*) Cristae extremely small and irregular as in the interstitial cells of *Opossum*.

(*viii*) Rarely the mitochondrial wall is smooth with no cirstae. The number and size of cristae in a mitochondrion directly affects its efficiency. The greater and larger are the cristae, the faster is the speed of oxidation reaction.

(*ix*) Perpendicular to the long axis of mitochondria.

## MITOCHONDRIAL PARTICLES

According to earlier descriptions the outer surface to the outer membrane and the inner surface of the inner membrane were supposed to be covered with thousands of small particles. Those on the outer membrane were described as being stalk less and were called the *subunits of Parson*. There may be as many as 10,000 to 100,000 particles per mitochondrion. The stalked inner membrane particles were called the *subunits of Fernandez-Moran*, *Elmentary particles*, *$F_1$ particles* or the *oxiosomes* or *ETP* or *electron transport particles*. These particles are about 85Å in diameter and are regularly spaced at intervals of 10 nm on the inner membrane. There may be as many as $10^4$ to $10^5$ elementary particles per mitochondrion.

## ISOLATION OF MITOCHONDRIA

Mitochondria can be isolated from the cell in the living form for their physiological studies. The cell first treated with deoxycholate for their break down. Then they are passed in sucrose solution. The homogenate should be centrifuged for 10 minutes at the speed of 600 X g. From this homogenate upper substance is centrifuged at the speed of 8500 X g for 10 minutes. After this centrifugation, the upper microsomal fraction is discarded while the lower fraction consists of mitochondria and other particles like lysosomes.

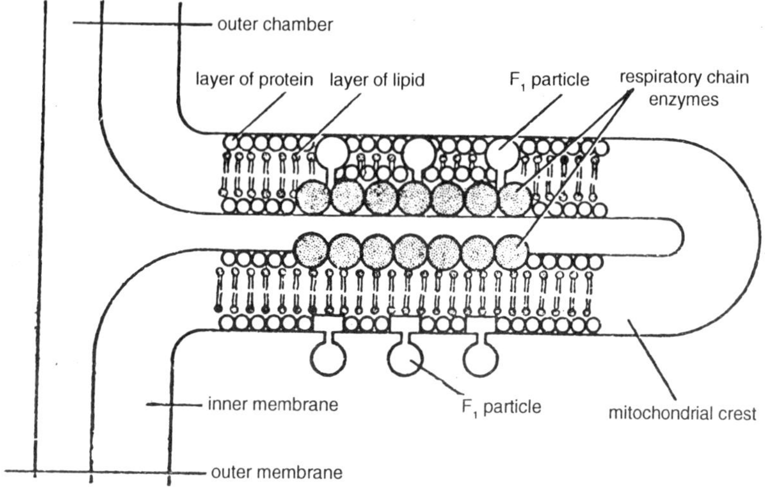


*Figure 4.7: Ultrastructure of a mitochondrial erest.*

This fraction is passed through sucrose gradient. The mitochondrial fraction then centrifuged at the speed of 10,000 X g upto 3 hrs. The upper part of this centrifuged material *have* mitochondria and lower part lysosomes.

## ISOLATION OF SUBMITOCHONDRIAL COMPONENTS

The complex structural organisation of the mitochondrion raises many questions concerning the functional significance of the various components that make up this organelle. The development of methods for separating and isolating these mitochondrial components has played an important role in advancing our knowledge in this area.

The first successful technique for separating inner and outer mitochondrial membranes was developed by Donald Parsons and his colleagues in the 1960s. In this procedure mitochondria are placed in a hypotonic solution until the outer membrane ruptures, and the inner outer Membranes are then separated from each other by isodensity centrifugation.

These two fractions can be readily distinguished *from* one another by electron microscopy; the isolated outer membranes look like empty sacs, while the isolated inner membranes form vesicles, called *mitoplasts,* which contain trapped matrix material within them.

The detergent digitonin has also been useful in isolating submitochondrial fractions because it selectively disrupts the outer mitochondrial membrane and thus allows the outer and inner membranes to be separated from each other by centrifugation. In both of the above procedures the contents of the intermembrane space are released into solution when the outer membrane is disrupted.

Hence any material appearing in the supernatant fraction after the initial centrifugation can be ascribed to the intermembrane space. Once the mitoplast fraction has been isolated, it can be further separated into its membrane and matrix components by treating it with the detergent Lubrol, which disrupts the inner membrane, and recentrifuging the resulting mixture.

Using this combination of techniques, the four major components of the mitochondrion can be separated from each other for biochemical analysis.

The highly permeable nature of the outer mitochondrial rane causes the solute composition of the intermembrane space to closely mirror that of the cell sap. The number of enzymes present in the intermembrane space appears to be relatively small. Prominent among these is adenylate kinase, an enzyme that catalyses transfer of the terminal phosphate group of ATP to AMP, forming two molecules of ADP.

## THE MITOCHONDRIAL MATRIX

The matrix space contains all the enzymes and cofactors involved in the Krebs cycle with the single exception of succinate debydrogenase, which is located in the inner membrane because it catalyses the direct transfer of electrons from the Krebs cycle intermediate, succinate, to the electron transfer chain.

Also present in the matrix are pyruvate dehydrogenase (which catalyses the conversion of pyruvate to acetyl-CoA), and the enzymes involved in fatty acid β-oxidation, which degrade fatty acids into acetyl-CoA units that enter the Krebs cycle. Analyses carried out on isolated matrix preparations have revealed that the protein concentration is too high for all the matrix proteins to be in true solution.

It has therefore been proposed that the enzymes of the Krebs cycle and fatty acid β-oxidation are anchored within a structural framework. This idea has received support from experiments carried out on mitochondria that have been artificially swollen by suspending them in a hypotonic medium. Under such conditions the matrix does not randomly disperse, but instead openings form in what appears to be an organised network. In electron micrographs the matrix generally exhibits a finely granular appearance, although large matrix granules ranging from 30 to several hundred manometers in diameter also occur.

In addition DNA, RNA, ribosomes, and other enzymes and factors involved in nucleic acid and protein synthesis are present in the matrix. Since those components are all involved in the process of mitochondrial growth and division, we shall delay their further consideration until the chapter on mitochondrial biogenesis.

## BIOCHEMISTRY OF MITOCHONDRIA

*Lindberg* and *Ernster* have given the data of chemical composition of mitochondria as follow : proteins 70 to 75% lipids 25-30%, and RNA 5% of the dry weight. But the cent biochemical analysis shows the following components

(i) *Proteins.* The proteins are the main constituent which insoluable in water. The outer limiting membrane of mitoiondria contains less than 10 percent of the total protein. sere are about 14 different proteins having molecular weight om 12,000 to 22,000. The inner membrane contains about )% protein having molecular weight varying from 10,000 to ),000. The protein composition of mitochondrial membranes not fully known.

(ii) *Lipid.* The lipid forms about 1/5th of the weight of the embranes. It is present almost entirely in the form of the olecules known as *phospholipid.* It has been reported by *Meluick* and *Packer* in 1971 that the outer membrane fraction is a 40% lipid contents as compared to 20% in the inner embrane.

(iii) *Enzymes.* About 70 enzymes and 12 co-enzymes have been recognised in the mitochondria. Enzymes lie in a nonlueours membranes region as solid arrays, with perhaps as any as 5000 to 20,000 such assemblages in a single liver or heart mitochondria.

### Mitochondrial DNA

Recently the DNA is also reported from mitochondria. he mitochondrial DNA is double stranded like the nuclear 1NA. Each mitochondrion may contain one or more DNA iolecules depending on its size, if, the mitochondrion is larger ten that may have more DNA molecules, having a circular shape.

Mitochondrial DNA differs from nuclear DNA in several aspects

The GC content is higher in mitochondrial DNA, id consequently the buoyant density is also higher. Another ifferences is the higher denaturation temperature of mitoiondrial DNA and facility with which it denatures.

The mount of genetic information carried by mitochondrial DNA not sufficient to provide specifications for all the proteins and enzymes present in this organoid. The most likely possibility is that mitochondrial DNA codes for some structural proteins.

**Table 4.2: Localisation of enzymes, obtained in fractionation studies.**

| *Mitochondrial fraction* | *Enzymes located* |
|---|---|
| 1. Outer membrane. | Monoamine oxidase, "Rotenone-insensitive" NADH- cytochrome C reductase, Kynurenine hydroxylase, Fatty acid CoA ligase, Glycerolphosphate acyl transferase, Nucleoside diphosphokinase, |
| 2. Inter-membrane space. | Adenylate kinase, Nucleoside diphosphokinase, Nucleoside monophosphokinase, |
| 3. Inner membrane. | Respiratory chain enzymes. P.Hydroxybutyrate, dehydrogenase, Ferrochelatase, Camitine palm ityl-transferase, Fatty acid oxidation system, Xylitol dehydrogenase, |
| 4. Matrix | Malate, isocitrate and glutamate dehydrogenase, Fumarase, Aconitase, Citrate synthetase, Omithine-Carbomyl transferase, Fatty acid oxidation systems, Pyruvate carboxylase, |

*Yeast* mitochondria have been shown to contain DNA polymerase and more recently *Kalf* has succeeded in isolating the enzymes from rat liver mitochondria. Mitochondrial DNA polymerase appears to be involved in DNA replication rather than repair and possesses properties which are different from those of the nuclear enzymes.

These include a differing requirement for metal ion. *Yeast* mitochondria) DNA polymerase appears to be smaller than its nuclear counter part, and is active at different stages of the cell cycle. Visual evidence showing what appears to be rat liver mitochondrial DNA in the process

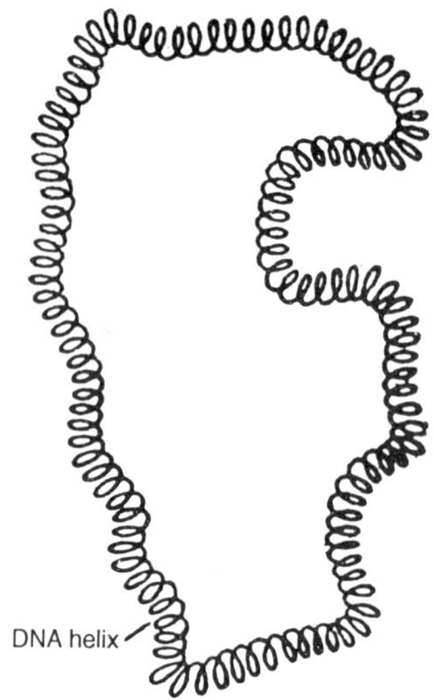


*Figure 4.8: Mitochondrial DNA.*

of replication has been presented by *Kirschner, Wolsten holme* and *Gross*.

Mitochondrial DNA does not appear to have histones associated with it as has nuclear DNA of higher organisms. In this respect mitochondrial DNA resembles with the bacterial DNA.

**Mitochondrial RNA**

*South* and *Mehlar* suggested that the amount of mt-RNA is about 10 to 20 times that of mt-DNA. All sorts of RNA have been identified in mitochondria. The present evidences point out conclusively the mitochondria contain complete set of t-RNA, aminoacyl RNA synthetases, as well as ribosomal RNA. All these components differ from their respective counterparts in the groundplasm.

The presence of m-RNA, transcribed from mitochondrial DNA is still uncertain. However, there are authorities who suggests its presence. The ribosomal RNAs' are coded for by mitochondrial DNA and are thus apparently synthesized within the mitochondria by a mitochondrial DNA-dependent RNA polymerase system.

**Mitochondrial Ribosome**

Mitochondria appear to contain ribosomes which are smaller in diameter than cytoplasmic ribosomes and *Yeast* mitochondria contain RNA species of 23S and 16S which would correspond to a 70S ribosome of bacterial type, than rather the 80S ribosome of the cytoplasm,

Ribosome like particles with sedimentation values of 81S and 55S have also been reported, and the extent of degradation suffered by the particles during isolation is not yet clear.

Polysome like aggregation of ribosomes have been observed in sections of yeast mitochondria by *Vignais, Huet* and *Andre* in 1969. High molecular weight and RNA species associated with mitochondria which differ in sedimentation value from cytoplasmic ribosomal RNA have been reported in Yeast, *Neurospora* and *He-La cells.* Mitochondrial ribosomes require a higher concentration of $Mg^{++}$ ions to maintain their integrity than do cytoplasmic ribosomes.

**Protein Synthesis**

In general, mitochondria can code any synthesize protein, but the DNA present therein is unsufficient to code for all the proteins. It is suggested that mitochondria can synthesize the proteins of structural nature (cytochrome oxidase), but much of the proteins if not all, of the soluble proteins of the matrix as well as the proteins of the outer membrane and a number of proteins located in the cristae are under the control of nuclear DNA.

Of the proteins coded by nuclear DNA, it is generally agreed that the m-RNA derived from nucleus are translated in the cytoplasm, and the resulting proteins are then transported into the mitochondria. How then protein enter the mitochondria ? Two methods have been proposed

(1) The precursors enter the mitochondrion and inside are changed into the end products, thereby affecting a undirectional flow a material into the mitochondria.

(2) There is the synthesis of lipoprotein vesicles which then merge and combine with the growing mitochondrion.

## FUNCTIONS

**Role of Mitochondria in Yolk Formation**

There has been a good number of investigations, whose account reveals that, mitochondria help in the formation of yolk in a developing ovun.. The first study in this field was made by *Loyez* and latest probably by *M.D.L. Srivastava*, with the help of light microscope. The evidence adduced depends upon the topographical, and size relationship, and staining reacticns of mitochondria and early protein yolk.

In the modern cytology with the investigation of electron microscope a new era has started and the studies of yolk formation do not remain away from electron microscope. With the help of electron microscope

*Farvard* and *Carasso* come to the conclusion that mitochondria transformed into yolk granules in the egg of *Planorbis coneus.*

The main structural changes which they observed in the mitochondria are as follows

(*i*) The cristae become disorganised in a few membranes, remaining concentric to the outer membrane before dropping completely.

(*ii*) In the matrix appeared a few minute granules which have scattered first, but become aggregated eventually in masses in regular pattern.

**During Cell division and Supermiogenesis**

Early cytologists, *Benda, Dueberg* and *Meves* were of opinion that mitochondria also divide equally during the cytoplasmic division and perhaps playing a part in inheritance.

*Wilson* commented that not the slightest proof has been produced of a fusion between the paternal and maternal chondrisomes. *Frederic* has briefly summarised various changes in mitochondria during cell division. The first phase shows a decrease in the total volume of mitochondrial material;. gradually ceases its movements, pronounced thinning, fragmentation into small spheres, loss of optical density and finally assimilation into the cytoplasm.

In the second phase, when the cell divides into two, the modified mitochondria are separated passively, into the daughter cells. In the third phase, the modified mitochondria are reconstituted by addition of elements assimilated in the cytoplasm. *Wilson* found that in *Opisthacanthus,* during spermatogenesis, the number of mitochondria gradually reduces.

*Pollister* describes in *Gerris* that mitochondria arranged themselves into a well defind ring, but without fusion. Modern microscopic studies provide firm conclusions regarding mitochondrial division during mitosis. *Pyne* noted in fowl adrenal cortex the mitochondria frequently appeared as pairs. This suggested that division, rather than fusion, was occurring.

In the transformation of spermatids into spermatozoa many ritochondrial changes are observed. *Franzen* observed in those sperms, which are shed directly in water, that mitochondria are present in form generally of four or five spheres below the sperm head, and in the case of sperms discharged in viscous medium these spheres transform into two elongated ribbion like filamentous mitochondria. Sometimes these develop into *'nebenkern spheres'* which may elongate and twist around the axial filament to form the mitochondrial sheath. *Yasuzumi* found an electron opaque body within 'nebenkern' of ;permatids of *Drosophilla ;* it is described as indistinguishable from a lipid droplet.

**Role of Mitochondria in Cellular Oxidation**

The carbon chains oxidised in cellular respiration originate from carbohydrates, fats, and proteins. As a preliminary to cellular oxidation, these three types of molecules are first converted into smaller units. Polysaccharides are hydrolysed into individual monosaccharides, fats are hydrolysed into glycerol and fatty acids, and proteins are split into amino acids. These products monosaccharides, glycerol, fatty acids, and amino acids-are the immediate fuels for cellular oxidation.

These fuels arc oxidised in a series of reactions that may include one or both of two major stages, the first occurring outside mitochondria and the second inside. In the first stage, called *glycolysis,* fuels are partially oxidised and converted to shorter three-carbon segments, which are completely oxidised to $CO_2$ and water in the second stage. Most of the ATP generated in cellular oxidation is produced in the second stage.

Monosaccharides and other carbohydrate units follow a main line of oxidation through both stages. Glycerol enters the glvcolytic pathway in the first stage, and the fatty acids enter mitochondria for oxidation in the second stage. Amino acids are deaminated (the amino, or —$NH_2$, group is removed) and converted into molecules that may enter the pathway in either stage.

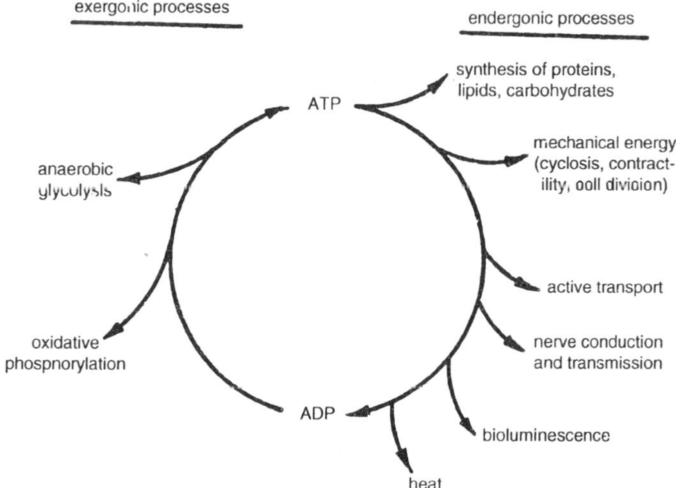


*Figure 4.9: Diagrammatic representation of uses and synthesis of ATP.*

### A Note about Oxidation

In the following discussion, it is important to remember that oxidation does not require direct combination of molecular oxygen with a metabolite. Oxidation describes any reaction in which electrons are removed from a molecule. For every oxidation, an electron acceptor is reduced by combining with the removed electrons. Therefore, each oxidation is accompanied by a reduction. Frequently oxidation involves removal of one or two hydrogen ions (protons) as well as electrons. The electron acceptors reduced during cellular oxidation may also combine with one or both of these hydrogens as well as with the electrons removed.

The amount of energy associated with the removed electrons depends on the orbitals they occupied in the oxidised molecule. Some of this energy is used to drive ATP synthesis in glycolysis and mitochondrial oxidations. The energy of the removed electrons can be expressed as a relative *potential* or *voltage* by comparison with an arbitrary standard. The standard used is the characteristic energy of electrons removed from hydrogen in the reaction $H_2 \rightarrow 2H^+ + 2e^-$ catalysed by platinum. The potential, or voltage, assigned to these electrons is 0.00 V. All other potentials, called *redox* or *reduction-oxidation potentials,* are measured and assigned a value with respect to this standard. The relative potentials of the electrons removed from me important intermediates in cellular oxidation.

### The First Stage of Cellular Oxidation : Glycolysis

The overall reactions of glycolysis split glucose and other six-carbon monosaccharides into three-carbon fragments. During glycolysis a single oxidation occurs, and each molecule of glucose yields a net gain of two ATP. Glycolysis has two major parts. In the first part, six-carbon molecules derived from glucose are raised to higher energy levels at the expense of ATP.

This part of the sequence in effect raises the derivatives of glucose to energy levels high enough to enter the second part. In the second part, the energy expended in the first part is recovered with a net gain in ATP through oxidation of the high energy derivatives of glucose. In the process, the glucose derivatives are split into two three-carbon units of *pyruvic acid.*

Each step in the glycolytic pathway is catalyzed by a specific enzyme, and the entire series of reactions involves 1 I enzymatic protein. The various enzymes, reactants, and products of glycolysis are in solution in the cytoplasm outside mitochondria and apparently interact through random collisions.

## The Reactions of the Glycolytic Sequence

The individual reactions in the glycolytic sequence and the enzymes catalysing them. In the first three reactions of the pathway, glucose is converted into a more reactive derivative ;ontaining two phosphate groups. Two molecules of ATP break down to provide the energy required for the phosphory ation of glucose.

The activity of the enzyme catalysing the irst reaction of the pathway, *hexokinase,* illustrates one of the nany co,itrols regulating the rate of oxidation in cells. The )roduct of the first reaction, *glucose-6-phosphate,* is an inhibitor of hexokinase. If glucose- 6-phospbate accumulates because be remainder of the sequence is running slowly, the hexokinase enzyme is inhibited, blocking further entry of glucose into the pathway.

Similar regulatory mechanisms control glycolysis as later points in the sequence. The glucose-6-phosphate produced in the first step is *rearranged* and then phosphorylated *into fructose-1, 6-diphosphate* at the expense of a second ATP. The characteristics of this reaction illustrate another mechanism regulating glycolysis, this one linked to the ATP supply.

The enzyme catalysing the phospborylation, *phosphofructokinase*, is inhibited by high concentrations of ATP and is stimulated by ADP and inorganic phosphate. If sufficient ATP is present in the cytoplasm, phosphofructokinase is inhibited and the subsequent reactions of glycolysis slow or stop. A surplus of other products of oxidative metabolism, such as reduced NAD, also inhibit the enzyme. If energy-requiring activities like place elsewhere in the cell, resulting in the conversion of ATP to ADP and phosphate, the accumulation of ADP and inorganic phosphate stimulates phosphofructokinase, increasing the rate of glycolysis and ATP production.

Regulation by phosphofructokinase is probably the most sensitive and significant of the controls of glycolysis, since it is directly keyed to the relative concentrations of ADP and ATP and thus to the late at which cells use energy for their activities.

The remaining reactions of glycolysis proceed without ATP. Fructose-1, 6-diphosphate is broken into two different three-carbon segments, which are readily interconverted, forming a "*pool*" of the two three-carbon sugars. One of these sugars, *3-phosphoglyceraldehyde (3-PGAL)* enters into the remaining steps in glycolysis. As 3-PGAL is depleted from the pool, it is replaced by conversion of the second tbreecarbon product.

At the next step in the sequence, two electrons and two hydrogens are removed from 3-PGAL. This oxidation is the primary energy-releasing step of glycolysis. Some of the energy released is trapped, as a part of the reaction, by the addition of a second phosphate from the medium (not from ATP) to form the product 1, 3-*diphosphoglyceric acid.* The electrons removed at this step have a relatively high potential or voltage and are accepted by a molecule called *nicotinamide adenine dinucleotide.*

NAD is a carrier molecule transporting highenergy electrons between cellular reaction system. The reduced NAD yielded at this step is one of the important high-energy products of the glycolytic sequence. The 1, 3-diphosphoglyceric acid product can also be regarded as a highenergy molecule, since removal of its two phosphates later in the sequence releases large increments of free energy, much of which is captured in the conversion of ADP to ATP.

The reaction removing the first of the phosphates is worth discussing in some detail, because it illustrates one method by which the energy of oxidation is converted into a usable chemical form in the cell. If 1 mol of 1, 3-diphosphoglyceric acid is hydrolysed directly to the products *3 phosphoglyceric acid* and inorganic phosphate, the reaction releases about 10,000 to 15,000 cal under standard conditions of temperature and pressure. In glycolysis, the breakdown of the diphosphorylated sugar is coupled to ATP synthesis.

The difference in calories between the two reactions represents rgy captured in the formation of ATP. The 3-phosphoglyceric acid product then enters a series of reactions, including a rearrangement and a conversion in which the elements of a molecule of water are removed. In the final reaction of glycolysis, the remaining phosphate is trans. ferred to ADP to produce ATP. The final reaction yields the end product of glycolysis, *pyruvic acid,* as well as ATP. Summarises the reactions of glycolysis.

Since each glucose molecule entering the pathway ultimately produces two molecules of 3-PGAL, a total of four molecules of ATP are produced in the conversion of 3-PGAL to pyruvic acid. The reactions attaching two phosphate to a glucose molecule in the initial steps require two molecules of ATP. Therefore, there is a net gain of two ATP, as well as two molecules of reduced NAD, for every glucose entering glycolysis. The overall reactants and products of glycolysis are therefore

$$\text{glucose} + 2\text{ADP} + 2\text{HPO}_4^{2-} + 2\text{NAD}_{ox} \rightarrow$$
$$2 \text{ pyruvic acid} + 2\text{NAD}_{red} + 2\text{ATP}$$

In addition to a pair of high-energy electrons, each reduced NAD molecule carries one of the two hydrogens removed in the oxidation of 3-phosphoglyceraldehyde. The other hydrogen enters the pool of $H^+$ ions in the medium surrounding the reaction sequence.

The two molecules of pyruvic acid yielded as final products are still relatively complex, energy-rich molecules that can be further oxidised to provide additional fret energy. These three-carbon molecules are the primary fuel foi the second major series of cellular oxidations that take place in mitochondria.

**Important Variations of the Basic Glycolytic Pathway**

The reactions of glycolysis from a central pathway in ti metabolism of carbohydrates in both plant and animal cell Starch in plants and glycogen in animals are both long-chai polysaccharides made up from repeating glucose links the enter the glycolytic pathway after hydrolysis into individu; glucose units.

Other sugars, including a wide variety of monc and-disaccharides, enter glycolysis after being converted 1: enzymes into one of the initial molecules of the pathway. At the opposite end of the sequence, pyruvic acid may be modified to yield other products. In one of the most important of these modifications, pyruvic acid is converted into *lactic acid* after accepting electrons from the NAD reduced earlier in the sequence.

pyruvic acid + reduced NAD → lactic acid + oxidised NAD

The most significant feature of this modification is regeneration of oxidised NAD, which is then free to cycle back to accept electrons in the oxidation of 3-PGAL in glycolysis. Because oxidised NAD is continually regenerated by this alternate pathway, glycolysis can continue to run with the net production of ATP.

This pathway is vital to cells living temporarily or permanently without oxygen (reduced NAD normally transfers its electrons through a series of carriers to oxygen ; see below). This pathway occurs in the muscle cells of animals, including man, if intensive, sustained physical activity is carried out before increases in breathing and heart rate have a chance to meet the demand for oxygen in the muscle tissue.

The lactic acid accumulating as a by-product is oxidised later, hen the oxygen content of the muscle cells returns to normal levels.

Another glycolytic variation occurs in organisms such as yeasts. In this modification, pyruvic acid accepts electrons From reduced NAD and is converted by additional reactions into *ethyl alcohol* (two carbons) and $CO_2$:

pyruvic acid+reduced NAD →

ethyl alcohol+$CO_2$+oxidised NAD

This variation in the glycolytic pathway is of central importance in human economics. It forms the biological basis of brewing and baking, and provides the alcohol of importance to one industry, and the $CO_2$ of significance to the other.

These variations in the glycolytic pathway, in which the electrons carried by reduced NAD are traded off to an *organic* substance such as pyruvic acid are collectively called *fermentations*. In the alternate pathway, the electrons carried from glycolysis by reduced NAD eventually reach an *inorganic* substance, molecular oxygen Fermentations of various kinds, producing a wide variety of products, are used as an ATP source by many species of bacteria.

Some of these species, called *strict anaerobes,* are limited to glycolytic fermentations for ATP production and cannot use oxygen at any time as a final electron acceptor. Others can use either glycolytic fermentations alone or both glycolysis and reactions equivalent to mitochondrial oxidations if oxygen is available. Bacteria in this category are called *facultative anaerobes*.

A number of species, termed *strict aerobes,* are unable to live by fermentation alone. Many cells of higher organisms, including the muscle cells of vertebrates described above, are facultative and can switch between fermentation and complete oxidation depending on their oxygen supply. Others are strict aerobes.

Many of the intermediate reactions of glycolysis are reversible and may go in either direction in response to high concentrations of either reactants or products. The reversible steps are indicated by those that are essentially irreversible are indicated by Alternate enzymatic pathways are available in the cytoplasm for the irreversible steps, enabling glycolysis to run in reverse.

For example, the first reaction in glycolysis, the conversion of glucose to glucose-6-phosphate by hexokinase, is essentially irreversible. However, an alternate enzyme, glucose-6-phosphatase, can catalyse the reverse reaction. Reversal of the pathway is an important part of reactions synthesizing six-carbon sugars and starch in both plant and animal cells.

### The Second Stage of Cellular Oxidations: Oxidation and ATP Synthesis in Mitochondria

In the second major stage of cellular oxidations, pyruvic acid, three-carbon molecule produced by the final steps of glycolysis, is completely oxidised to $CO_2$ and $H_2O$, releasing large quantities of free energy. The yield of ATP in these. reactions far exceeds the amounts obtained from glydolysis. The reactions occur in three parts, all located in mitochondria. In the first, pyruvic acid is shortened into a two-carbon segment, and one molecule of $CO_2$ is released.

In the second the two carbon segment is completely oxidised to two molecules of $CO_2$. In the third part, the electrons removed in these oxidations travel through a series of electron carriers to reach oxygen. Much of the free energy released by this transfer is used to drive the synthesis of ATP. Because oxygen is the final electron acceptor for these reactions the oxidative activities of mitochondria are frequently termed *respiration.*

**Part I of Mitochondrial Respiration :**
**Oxidation of Pyruvic Acid to two-Carbon Segments**

After entering mitochondria, pyruvic acid is shortened to :wo-carbon segment in a cyclic series of reactions. These fictions remove two electrons, two hydrogens, and one carbon $CO_2$) from the pyruvic acid chain. The two-carbon seg:nt produced by the oxidation is an *acetyl* ($-CH_3CO^-$) group:

$$\text{pyruvic acid- acetyl group} + 2e^- + 2H^+ + CO_2$$

The two electrons and one of the hydrogens removed in this ion are transferred to NAD. The $CO_2$ and the remaining hydrogen are released to enter the surrounding medium. The two-carbon acetyl group is transferred to an acceptor called *coenzyme A* to produce the high-energy substance *acttyt coenzyme A*. Coenzyme A is another carrier molecule based on nucleotide structure, as are ATP and NAD.

Coenzyme A accepts and carries two-carbon acetyl units, as ATP carries phosphates and NAD transports electrons. Most of the free energy released in the oxidation of pyruvic acid is captured as chemical energy in the conversion of coenzyme A to acetyl coenzyme A.

The overall reaction cycle in pyruvic acid oxidation. thus yields as net products acetyl coenzyme A, reduced NAD, and $CO_2$:

$$\text{pyruvic acid} + \text{oxidised NAD} + \text{coenzyme} \rightarrow$$
$$\text{acetyl co-enzyme A} + \text{reduced NAD} + CO_2$$

All of the reactants and products in Reaction 7-6 are multiplied by a factor of 2 if pyruvic acid oxidation is considered as a continuation of glucose oxidation, since each molecule of glucose entering glycolysis produces two molecules of pyruvic acid.

Pyruvic acid oxidation is catalysed by a molecular aggregate containing five enzymes and a series of intermediate electron and hydrogen carriers. The reactions carried out by this multienzyme complex, worked out primarily in the laboratory of L.J. Reed, proceed. The enzyme carrying out the first steps of the cycle, *pyruvic acid dehydrogenase, is* another regulatory enzyme that tunes the rate of cellular oxidations to the cell's energy needs.

High ATP concentrations inhibit the activity of pyruvic acid dehydrogenase, slowing or stopping pyruvic acids oxidation. The enzyme is also inhibited, directly or indirectly, by high concentrations of reduced NAD or acetyl coenzyme *A*. High ADP concentrations, in contrast, stimulate activity of the enzyme. The result is a sensitive control mechanism that closely matches the rate of pyruvic acid oxidation to the energy requirements of the cell.

The enzyme-acceptor complex carrying out pyruvic acid oxidation exists in mitochondria in the form of aggregates containing as many as 100 or more polypeptide chains, including many copies of the central enzyme of the complex, pyruvicacid dehydrogenase.

Bacteria able to carry out oxidation of pyruvic acid possess a similar complex. A recent analysis ofpyruvic acid oxidation in *E. coli* indicates that as many as 60 polypeptide chains may be present in a single aggregate in this species, with a total molecular weight of 4,800,000.

Both organic products of pyruvic acid oxidation are highenergy substances. The acetyl units carried by acetyl coenzyme A serve as the immediate fuel for the remaining oxidative reactions of mitochondria, and the electrons carried by reduced NAD represent potential energy that is eventually tapped off to drive the synthesis of ATP.

## Part 2 Mitochondrial Respiration : Oxidation of Acetyl Units to $CO_2$ in Mitochondria

The acetyl group carried by coenzyme A is oxidised to $CO_2$ in a cyclic series of reactions first deduced in 1937 by a British investigator, Hans Krebs, who received the Nobel Prize for his brilliant work with cellular oxidation. In the cycle which is named for Krebs, there is a continuous input of two-carbon acetyl units as reactants and a continuous output of the products : $CO_2$, ATP, reduced NAD, and an additional electron carrier in reduced form, *flavin adenine dinucleotide.*

The molecules forming intermediate parts of the cycle are continuously regenerated as the cycle turns. Only one molecule of ATP is. formed as a direct product of each turn of the Krebs cycle.

Most of the energy released by the several oxidations of the cycle is trapped in the electrons carried from the cycle by reduced NAD and FAD.

The reactions of the Krebs cycle proceed. In the first step in the sequence, the acetate group linked to acetyl coenzyme A is transferred to *oxaloacetic acid* (four carbons), forming *citric acid (six* carbons). This reaction releases coenzyme A, which is then free to recycle through another oxidation of pyruvic. acid

acetyl coenzyme A+oxaloacetic acid—citric acid -+ coenzyme A The enzyme catalysing Reaction, *citrate synthetase, is* one of the major regulators of the Krebs cycle. The enzyme is inhibited and the first reaction of the cycle slows or stops if succinyl -coenzyme A, a product of a reaction later in the cycle, accumulates. The enzyme is also inhibited by high concentration of ATP.

The citric acid product of the first reaction is rearranged in a series of steps into *isocitric acid,* which becomes the reactant for the first oxidation of the cycle. This oxidation is catalysed by either of two enzymes both of which are called *isocitrate dehydrogenase.* The reactions catalysed by are enzymes are identical except that one uses NAD and the other the closely related substance *nicotinamide adenine dinucleotide phosphate as* the electron acceptor.

Two electrons and two hydrogens are removed from isocitric acid in the oxidation; at the same time, the carbon chain is reduced in length from six to five carbons, yielding *α-ketoglutaric acid* as product. The carbon removed is released as $CO_2$. Which of the two enzymes catalyse this reaction depends on the relative concentrations of ADP and ATP in the medium. The NAD-specific enzyme is more active than the NADP form when the ATP concentration is relatively low.

If ATP concentration is high, the NAD-specific enzyme is inhibited, and NADP is favoured. High concentration of reduced NAD, which have a similar inhibitory effect on the NAD-specific enzyme, also favour NADP as the electron acceptor. On the other hand, high ADP concentrations stimulate the activity of the AD-specific enzyme, which actually requires ADP as a cofactor for its activity.

These differences are significant for ATP production because reduced NAD normally transfers its electrons to the electron transport system of mitochondria that generates ATP. Reduced NADP instead acts as an electron donor for reactions that build up more complex substances in which reductions are required. As a result of this system, isocitric acid oxidation is finely turned to the needs of the cell for

ATP. Under the usual conditions, in which cellular activity demands a more or less continuous supply of ATP, the isocitric acid dehydrogenase enzyme using NAD as an electron acceptor predominates in Krebs cycle oxidation at this step, leading to ATP synthesis.

At the next step in the cycle, *α-ketoglutaric* acid is oxidised to *succinie acid* in a side cycle that closely resembles pyruvic acid oxidation. The enzyme-acceptor complex active in this reaction includes essentially the same types of enzymes and intermediate electron acceptors. The central enzyme in the complex, however, is *α-ketoglutaric acid dehydrogenase,* which catalyses removal of two electrons and two hydrogens from a-ketoglutaric acid. At the same time, one carbon is removed frog the acid and released as $CO_2$. The product, a succinyl group (four carbons) remains attached to the complex until its transfer to coenzyme A in a step analogous to reaction 3. This transfer forms the high-energy product *succinyl coenzyme A.*

The electrons and hydrogens are then transferred from acceptors of the complex to NAD, in a step equivalent to reaction 4. Coenzyme A is regenerated by a subsequent reaction in which the succinyl group is removed, releasing succinic acid and a large increment of free energy. Much of this free energy is captured in the synthesis of a molecule of ATP, the only ATP, originating directly from the Krebs cycle

$$\text{succinyl coenzyme A} + \text{ADP} + HPO_4^{2} \rightarrow$$
$$\text{succinic acid} + \text{coenzyme A} + \text{ATP}$$

ATP synthesis proceeds directly as shown in this reaction in bacteria. In animals, the enzyme catalysing the reaction phosphorylates guanosine diphosphate (GDP) instead. However, the guanosine triphosphate (GTP) produced is subsequently converted to ATP by a mitochondrial enzyme that readily interconverts the various nucleoside triphospbates

$$\text{GTP+ADP} \rightleftharpoons \text{ATP+GDP}$$

As net products, the overall reaction sequence oxidising a-ketoglutaric acid thus yields

$$\alpha\text{-ketoglutaric acid} + \text{oxidised NAD} + \text{ADP} + HPO_4^{2} \rightarrow$$
$$\text{succinic acid} + \text{reduced NAD} + \text{ATP} + CO_2$$

Succinic acid is oxidised at the next step of the cycle by the enzyme *succinic acid dehydrogenase.* The enzyme is unique mong the various proteins catalysing the Krebs cycle because t is tightly bound to the inner mitochondrial membranes, ,here it forms part of the integral proteins of the cristae. Also ound to the enzyeme in the membrane is its own electron cceptor FAD. FAD is another carrier molecule resembling ATP, NAD, NADP, and coenzyme A in structure.

The comination of this acceptor with the enzyme is known as *a flavorotein*. In the reaction catalysed by succinic acid dehydroenase, FAD is reduced by accepting two electrons and both iydrogens removed from succinic acid. The product of the eaction is the four-carbon molecule *fumaric acid*

succinic acid +oxidised FAD → fumaric acid + reduced FAD

The final oxidation of the Krebs cycle occurs after a rearranging reaction in which fumaric acid is converted to *malic icid* by the addition of the elements of a molecule of water. Malic acid is then oxidised to oxaloacetic acid. NAD acts as in electron acceptor for the reaction and also binds one of the two hydrogens removed from malic acid. The molecule of *oxaloacetic acid* produced replaces the oxaloacetic acid used in :he first reaction of the Krebs cycle, and the. cycle is ready to turn again, summarises the reactions of the cycle in simplified form.

**Products of the Krebs Cycle**

We can now determine the overall products of the Krebs ;ycle and summarise the cellular oxidation of glucose discussed ;o far. As the Krebs cycle proceeds through one complete turn, one two-carbon acetyl unit is consumed and two mole-ules of $CO_2$ are released. At each of four reactions in the cycle two electrons are removed. We will assume for this summary that the first oxidation of the cycle uses NAD rather than NADP as the electron acceptor.

At three of the oxida:ions, then, NAD is the acceptor, producing three reduced NAD molecules; one step produces a molecule of reduced FAD. As a part of the oxidation of α-ketoglutaric acid, one molecule of ATP is generated. Oxaloacetic acid, used in the initial reaction of the Krebs cycle, is regenerated in the final reaction. Thus, as overall reactants and products, the Krebs cycle includes

acetyl unit +3 oxidised NAD+ oxidised FAD+
$ADP+HPO_4^{2} \rightarrow 2CO_2$+3 reduced NAD+
reduced FAD+ATP

With this information we can sum up all of the products of complete oxidation of glucose to $CO_2$, from glycolysis through the Krebs cycle. For each molecule of glucose entering the series, the Krebs cycle will turn twice. Glycolysis and pyruvic acid oxidation together yield four reduced NAD, two ATP, and two $CO_2$ molecules, since two molecules of pyruvic acid are oxidised for each glucose molecule entering the sequence. Adding to this the products of two turns of the Krebs cycle for every glucose entering oxidation, we have:

glucose + 4ADP + 4HPO4$^{2-}$ + 10 oxidised NAD +
2 oxidised FAD $\rightarrow$ 4ATP+10 reduced NAD+
2 reduced FAD+6$CO_2$

Note that at this point little ATP has been produced. However, the electrons carried by the 10 reduced NAD and 2 reduced FAD molecules occupy orbitals at high-energy levels and contain most of the energy obtained from oxidation of the glucose to six $CO_2$ molecules. The free energy released in the transfer of these electrons to oxygen is used as the ' major source of energy for cellular ATP synthesis in the third part of mitochondrial oxidations.

**The Carrier Molecules**

With one exception, all of the known electron carriers consist of a protein molecule in combination with a tightly bound, nonprotein prosthetic group that is the actual electron carrier. The prosthetic groups are alternately oxidised and reduced as electrons flow through the system.

The single exceptional carrier, *coenzyme Q*, is a lipidlike substance suspended in the membrane interior without direct linkage to a protein. Although coenzyme Q and most of the prosthetic ;groups have been isolated and chemically characterised, the protein portions of the carriers, except cytochrome c, are integral membrane proteins that can be removed only by techniques that seriously denature them. As a result, the protein segments of the carriers are relatively poorly known. Four major kinds of carriers have been identified in the chain. The flavoproteins, including FAD and *flavin mononucleotide* (FMN) are electron carriers with prosthetic groups based on nucleotides.

In FAD and FMN, the part of the prosthetic groups alternately oxidised and reduced during electron transport is derived from *riboflavin,* a vitamin of the B group. Since this prosthetic group contains a nitrogenous base, a five-carbon sugar, and a phosphate group, it is classified as a nucleotide. The flavoproteins carry electrons in pairs and also bind two hydrogens of form $FADH_2$ or $FMNH_2$ during reduction.

The *cytochromes,* including cytochrome *a*, $a_3$, *b*, *c*, and *d* have a complex porphyrin ring containing a central iron atom as a prosthetic group. The various cytochromes differ in minor substitutions in side groups attached to the porphyrin ring. In each, alternation between the oxidised and reduced forms occurs by means of a single electron gained or lost by the central iron atom, which can exist as either $Fe^{2+}$ or $Fe^{3+}$. In contrast to the flavoproteins, the cytochromes carry no hydro gens. Cytochrome c is unique among the electron carriers in, that it is a peripheral membrane protein, easily removed from mitochondrial membranes without being denatured. As a result, cytochrome c is the best-known of the protein-based electron carriers.

*Iron-sulfur proteins* comprise the third kind of mitochordrial electron carrier. As yet, these proteins are poorly characterised and are known only to contain centers containing iron atoms in close association with-SH (thiol' groups. Presumably, the iron and sulfur atoms are part of a prosthetic group attached to the protein ; in reduction and oxidation, the iron atoms are believed to alternate between $Fe^{2+}$ and $Fe^{3+}$ states. At least seven different iron-sulfur proteins have been detected in the mitochondrial electron transport system. As far as is known, these carriers transport single electrons and ro. hydrogens, in a pattern similar to that of the cytochromes.

The final kind of electron carrier, *coenzyme* Q contains a quinone ring that is alternately oxidised and reduced during electron transport. Attached to the ring is a hydrophobic side chain, which in mitochondria consists of a series of ten repeating subunits. Coenzyme Q is a dual electron-hydrogen carrier and has two redox levels or states in which it may accept either one electron and hydrogen or electrons and hydrogens in pairs. This characteristic figures in one of the latest modifications of Peter Mitchell's cherniosmotic hypothesis.

**Reconstructing the Electron Carrier Sequence**

A variety of biochemical techniques has been used to assign the known carriers to their probable locations in the electron transport chain. Part of this work is based on the measured redox potentials of the purified carriers or prosthetic groups. Electrons released from the various carriers have a characteristic energy; by arranging the carriers in order of increasingly positive potential, their sequence in mitochondrial electron transport can be approximated.

This method suffers from the fact that redox potentials measured under standard conditions, when the carriers are in isolated form, may differ from the potentials in the natural situation in the membrane, leading to the possibility of errors in placing them in sequence. A somewhat more direct method for sequencing the carriers .s based on the fact that most of the carrier molecules undergo pronounced colour changes on conversion from oxidised to reduced form.

The flavoproteins, for example, are yellowish in solution when oxidised and colourless when reduced. These changes can be traced quantitatively by measuring the *absorption spectra* of the molecules in solution (or in mitochondrial membranes). In this technique, the amount of light absorbed by the carrier at each wavelength is measured and plotted. As colour changes occur on oxidation or reduction, characteristic absorption peaks appear and disappear in the spectrum plotted for the molecule.

These absorption spectra were used in the laboratory of critton Chance to place the carriers in a tentative sequence. Chance and his coworkers found that in intact, active mitochondria, most of the FAD or FMN molecules are in the reduced state. Two of the cytocbromes, a and $a_3$, tend to be present almost entirely in the oxidised state. Other carriers fell between these two extremes, falling into a natural sequence in terms of the ratio of oxidised to reduced forms.

Reasoning that the carriers appearing in highest proportion in the oxidised form were nearest oxygen in the sequence, Chance assigned each carrier a position in the electron transport system. This method is also not without pitfalls, since in intact mitochondria some of the absorption peaks of different carriers overlap, making interpretation of the spectra obtained difficult. Also, some of the carriers, such as coenzyme Q and the iron-sulfur proteins, cannot be readily detected or assigned definite positions by this technique.

These results were strengthened by Chance and others by combining the light absorption studies with the use of inhibitors known to combine with specific carriers and interfere with their oxidation or reduction. Carriers falling after the inhibited carrier in the sequence gradually become oxidised by removal of electrons and show the characteristic colours for the oxidised forms.

Carriers falling before the blocked carrier remain in the reduced form and show colours characteristic for this state. For example, the drug *antimycin a* blocks electron transport from cytochrome b. In mitochondria exposed to the drug, NAD, the flavoproteins, and cytochrome *b* remain in the reduced state, and cytochromes a and c become oxidised. This indicates that NAD and the flavoproteins fall before cytochrome *b,* while cytochromes *a* and c fall after it.

By using different inhibitors and nothing the "*crossover points*" between carriers converted to reduced or oxidised states, the tentative positions assigned by means of redox potentials and absorption spectra were strengthened. Combining the results of these approaches established the probable carrier sequence. Note that electrons have two routes of entry into the chain, one at FMN and one at FAD, which join at coenzyme Q. Electrons flow spontaneously along the chain, losing energy as they move from orbitals in one carrier to orbitals in the next. At some of the steps, sufficient energy is released to drive the synthesis of ATP. As noted, some of the carriers carry electrons in pairs, and some singly. Hydrogen combines directly only with NAD, FAD, FMN, and coenzyme Q and at the end of the sequence with oxygen.

When electrons pass from a hydrogen to a nonhydrogen carrier, hydrogen is considered to be released to the medium as $H^+$ ions or, in the case of the Mitchell hypothesis, to be expelled across the inner mitochondrial membranes. At the level of the final acceptor, hydrogen ions reenter the sequence in the reduction of oxygen to water.

Because of the ambiguities in the different sequencing techniques, there are still uncertainties about the placement of some of the carriers. These difficulties are most pronounced in the assignment of the iron-sulfur proteins. The total number o[ these carriers and their positions in the electron transport sequence are still controversial. Because of overlapping spectra in the light wavelengths absorbed by cytochrome *b*, it is unclear at present whether there are one, two, or three different molecules of this cytochrome transferring electrons between coenzyme Q and cytochrome $c_1$. Similarly, it is presently unknown whether the two prosthetic groups identified as cytochromes *a* and $a_3$ are carried on the same protein molecule or different proteins. Because *a* and $a_3$ are always closely associated, the consensus at the present time is that they combine into one molecule, presumably in the form of two porphyrin rings, one of the a and one of the $a_3$ type, carried in crevices in a single protein.

**Electron Transport and ATP Synthesis**

During the late 1940s, A. L. Lehninger and his associates established that transport of electrons along the mitochondrial carrier system, from reduced NAD to oxygen, is directly linked to ATP synthesis. Lehninger showed that isolated initochondria treated to mcke them permeable to NAD could synthesize ATP if placed in a medium containing reduced NAD, ADP, phosphate, and $Mg^{2+}$ ions. For each molecule of reduced NAD added, the isolated mitochondria synthesized three molecules of ATP.

When ascorbic acid, an electron donor incapable of reducing carriers higher in the sequence than cytochrome c, was added to the isolated mitochondria, only one phosphorylation of ADP to ATP occurred for each molecule of ascorbic acid. This showed that energy sufficient to drive the synthesis of one molecule of ATP is released somewhere in the sequence from cytochrome *c* to oxygen.

Other possible sites of significant energy release were located by an extension of the light absorption studies used to sequence the carriers. Chance and G.R. Williams used a mitochondrial preparation having a limited supply of ADP and an excess of oxidisable substrate. The limited supply of ADP caused bottlenecks at three sites in the electron transport system, which were detectable by colour changes indicating an excess of reduced NAD, cytochrome b, and cytochrome c. This indicated, for example, that the step in electron transfer from NAD to flavoprotein is closely coupled to the synthesis of one ATP molecule; a limited supply of ADP causes the accumulation of reduced NAD.

More recently, the findings from these studies have been supplemented by the results of studying isolated parts of the electron transport chain after disruption of mitochondrial membranes by sonication or detergents. By this approach, carried out by D.E. Green, E. Racker and his associates, and others the electron transport chain was broken into segments containing several carriers or into individual carrier molecules. Reassembling the carriers in various combinations with membrane segments and the ATP-synthesizing enzyme then allowed an evaluation of the capacity of parts of the electron transport system to drive ATP synthesis.

The combined results of these techniques indicate that ATP is synthesized in the mitochondrial transport chain as pairs of electrons pass from NAD to coenzyme Q (site I), from cytochrome *b* to cytochrome c (site II), and from cytochrome $a$-$a_2$ to oxygen. Thus, each pair traversing the transport system from reduced NAD to oxygen generates three ATP molecules. Since the electrons carried by FAD enter the chain at a point past site I, the first ATP generating site, the electrons carried by reduced FAD from succinic acid oxidation give rise to only two ATP molecules.

Synthesis of ATP in the isolated and reassembled transport systems occurs only if three basic conditions are met :

(1) The various carriers must be combined with phospholipids into bilayers or intact inner mitochondrial membranes;

(2) the membranes must be closed into sealed vesicles with an inside space completely separated by intact membranes from the medium outside the vesicles; and

(3) the mitochondrial ATP synthesizing enzyme must also be present in the sealed membranes of the vesicles with the carrier molecules. All of these conditions are satisfied in intact mitochondria.

**A Summary cf Glucose Metabolism**

With this information, the total number of ATP molecules synthesized for each molecule of glucose oxidised to $CO_2$ and $H_2O$ can be calculated. Complete oxidation of one molecule of glucose (through glycolysis, acetyl coenzyme A formation, and the citric acid cycle) results in a total of 4ATP, 10 reduced NAD, 2 reduced FAD, and $6CO_2$ molecules.

The ten electron pairs carried by reeuced NAD, traversing the electron transport system, will result in the synthesis of $10 \times 3 = 30$ molecules of ATP. Reduced ADF, carrying electrons that enter the transport system at a later point, will cause the synthesis of $2 \times 2 = 4$ molecules of ATP. Added to the total the overall oxidation of glucose yields 38 molecules of ATP

$$\text{glucose} + 38ADP + 38HPO_4^{2} + 6O_2 \rightarrow 6CO_2 + 44H_2O + 38ATP$$

This total of 38 ATP assumes that the two molecules of NAD reduced in glycolysis will each induce synthesis of three ATP molecules inside the mitochondrion. However, reduced NAD from glycolysis cannot directly enter the mitochondrion to pass electrons to the transport system because the mitochondrial membranes are impermeable to NAD. Instead, the electrons carried by reduced NAD outside mitochondria are transferred to other substances that act as electron shuttles between the mitochondrial exterior and interior.

Two of these shuttle mechanisms are known. One is more ficient and results in the transfer of electrons from NAD outde to NAD inside the mitochondrion. The second is less ficient and results in the reduction of FAD inside the itochondrion. If the more efficient shuttle is predominant, o significant energy loss occurs, and 38 ATP will result orn each molecule of glucose completely oxidised.

If the :ss efficient shuttle predominates, the electrons from reduced TAD are transferred instead to FAD inside the mitochonrion. These enter the electron transport system farther long in the chain and result in the synthesis of 2 ATP for ach molecule of reduced FAD. In this case, complete oxidaion of glucose will result in a total of 36 .ATP molecules nstead of 38. Which shuttle predominates depends on the articular species and tissues involved.

Under standard conditions (pH 7, reactants and products at 25°C and I M), the hydrolysis of ATP to ADP yields 7000 -al mol. Using this value as the energy required to synthesize ATP from ADP and $HPO_4^{2-}$, the total energy trapped during the oxidation of glucose may amount to either $36 \times 7000 = 252{,}000$ cal/mol or $38 \times 7000 = 266{,}000$ cal/

mol. Combution of glucose in air yieids 686,000 calfmol. On this basis, the efficiency of glucose metabolism in cells falls between 37% and 39% (252/686 × 100 and 266/686 × 100).

The energy captured under the temperature and concentration conditions in the living cell may actually be considerably higher than these figures indicate. The free energy change fo r the hydrolysis of ATP increases in magnitude at the concentration of the reactants and products is reduced. The concentration of ADP and ATP at the site of phosphorylation is not known with certainty.

Estimates of the actual free energy change vary from 9000 to 15,000 cal/mol depending on the concentrations assumed to occur in the cell. Using these figures gives a higher efficiency for the total oxidation of glucose, from about 50% at 9000 cal/mol to 80% at 15,000 cal/mol. Where the actual figure falls in this range is unknown, but the efficiency is probably higher than the 37-39% calculated from standard conditions. Even at the 37-39% level, the efficiency of mitochondrial energy conversion is considerably higher than that of most energy-conversion systems designed by human engineers, which rarely perform above the 5.10% efficiency level.

**ATP Synthesis in Mitochondria**

The research effort to unravel the mechanism coupling ATP synthesis to mitochondrial electron transport called *oxidative phosphorylation,* is presently among the most extensive and active in cell biology. Experimentation, hypothesis, and speculation in this field have concentrated on two areas : characterisation of the ATP-synthesizing enzyme and its mechanism of action, and determining how the energy released in electron transport is used to power ATP synthesis. The experiments and ideas connected with this field of research (by scientists who are sometimes called "*mitochondriacs*") are among the most elaborate in contemporary biology. Understanding them Fill require extra effort ; if you become confused, you are in good company. Efraim Racker, one of the foremost workers in the field, once said in a talk that "anyone who is not confu sed about oxidative pbosphorylation just does not understand the situation."

**The ATP-Synthesizing Enzyme of Mitochondrial Electron Transport**

Extraction and purification of mitochondrial membrane fractions reveals that the enzyme catalysing the synthesis of ATP in oxidative phosphorylation is embedded in the cristae membranes, in the same general location as the electron transport carrier molecules. Work by

A.E. Senior, E. Racker, and others has shown that the enzyme is contained in a complex consisting of ten or more polypeptides. Some of these polypeptides function to anchor the complex in the membrane, and some directly catalyse the reaction sythesizing ATP:

$$ADP + HPO_4^{2-} \rightleftharpoons ATP + H_2O$$

Racker's work has allowed morphological identification of the ATPase enzyme with a lollipop-shaped particle that under some conditions can be seen extending from the inner mitochondrial membrane into the mitochondrial matrix. In Racker's experiments with the particles, carried out with L.L. Horstman, mitochondria were isolated and mechanically disrupted by sonication or shaking in a medium containing glass beads. This treatment ruptures both the outer and inner mitochondrial (cristae) membranes. The reptured inner membranes seal spontaneously into closed vesicles capable of electron transport and ATP synthesis.

The inner membrane preparation, if viewed in the electron microscope after negative staining, shows the typical lollipop particles. Racker and Horstman treated isolated membranes of this type by exposing them to urea, a chemical that disrupts hydrogen bonds. After this treatment, the membranes were still capable of electron transport but could not synthesize ATP.

Examination after negative staining showed that the lollipop particles were missing from the membrane surfaces. The removed particles, if separately purified, were found to be capable of ATP hydrolysis, that is, of running Reaction in reverse. Thus, the removed particles contained the ATPase activity associated with the inner membranes in intact mitochondria. Adding the purified particles back to the membrane preparations restored both the lollipops and the capacity for electron-transport-driven ATP synthesis to the membrane preparation.

Analysis of the inner membrane particles reveals that the ATPase activity occurs in the spherical "headpiece" of the lollipop. At least five different polypeptides can be identified with the headpiece, which is connected to the membrane by a "*stalk*" containing at least two polypeptides. Anchoring the headpiece and stalk to the membrane is the "*base*," containing an additional three or four hydrophobic polypeptides normally embedded deeply in the membrane.

The headpiece is easily removed from the stalk and base by various mild treatments, such as exposure to urea or low salt concentrations, indicating that in intact membranes, it is a peripheral protein. The lollipop structures are made visible most readily by negative staining. The particles are rarely seen in thin-sectioned mitochondria or in freeze-fracture preparations. As a result, it is not certain whether the particles actually extend from the inner membrane surfaces in intact mitochondria or are somehow extruded from a normal location closer to the membrane surface by the isolation or staining technique.

**The Mechanism Coupling ATP Synthesis to Electron Transport**

In intact mithchondria, ATP synthesis is tightly coupled to electron transport. If electrons move through the system, ATP is synthesized ; if no ADP is available, thus blocking ATP synthesis, electrons cannot pass along the carriers from NAD or FAD to oxygen. The nature of the coupling mechanism has been debated for many years and is still the subject of much controversy. Two major hypotheses are now considered seriously.

One is the chemiosmotic hypothesis of Peter Mitchell. which has steadily gained experimental support from its first statement in 1961 until the present ; its basic tenets have been accepted by most investigators in the field. The second model, the *conformational hypothesis,* first advanced by P.D. Boyer in 1963, has gained recent attention because it can account for certain details not adequately explained by the chemiosmotic hypothesis.

**The Chemiosmotic Hypothesis**

Mitchell's hypothesis proposes that mitochondrial ATP synthesis is driven by a gradient of $H^+$ ions set up directly by electron transport. According to Mitchell, $H^+$ ions are expelled to one side of the membrane as electrons travel from NAD or FAD to oxygen. The $H^+$ gradient established by the electron transport acts as a source of free energy that drives the synthesis of ATP from ADP and $HPO_4^{2-}$ as the gradient runs down.

***Establishing the $H^+$ gradient***

The $H^+$ gradient, according to the most recent modifications of the chemiosmotic hypothesis, depends on the fact that some of the carriers in the electron transport chain carry both hydrogens and electrons during their cycles of oxidation and reduction, while some carry electrons only.

In the first steps in the mechanism, reduced NAD (NADH) passes its two electrons and one hydrogen to FMN. Since FMN carries hydrogens in addition to an electron pair, a second $H^+$ is absorbed from the medium inside the matrix to form $FMNH_2$. This carrier passes its electrons to the next carrier group in the chain, the iron-sulfur proteins. Because the iron-sulfur proteins are "pure" electron carriers, the hrdrogens carried by $FMNH_2$ are released. Release occurs into the intermembrane compartment, contributing the first two $H^+$ ions to the gradient.

The next series of steps involves coenzyme Q in a mechanism that Mitchell calls the *Q cycle*. In the first step of the Q cycle, the two electrons carried by the iron-sulfur proteins are passed to two molecules of coenzyme Q, one to each Q molecule. These take up one H+ each from the matrix at the same time and go to the QH, or semiquinone form. To go to the fully reduced, or $QH_2$, form the two Q molecules each accept an electron from a cytochrome *b* molecule (the source of the electrons carried by the two cytochrome *b* molecules required for this transfer will soon be apparent'.

The additional pair of $H^+$ ions required is taken up from the matrix. Thus, in going from Q to $QH_2$, the two coenzyme Q molecules absorb 4H from the matrix. In the next step in the Q cycle, the two coenzyme Q molecules pass one electron each to the next carriers in the chain, two molecules of cytochrome c. Since the cytochromes are pure electron carries, the coenzyme Q molecules, in going from $QH_2$ to QH, release one $H^+$ each to the intermembrane compartment.

The second electron carried by each QH molecule, according to Mitchell, passes to each of two cytochrome *b* molecules. As this transfer takes place, the two coenzyme Q molecules go to the fully oxidised or Q form and release their last $H^+$ ions, one each, to the intermembrane compartment. A total of $6H^+$ is thus transferred from the matrix to the intermembrane compartment as a pair of electrons passes from reduced NAD to this point in the chain.

The electrons carried by cytochrome c then move along the chain through *a*-$a_3$ cytochromes to oxygen. As the electron pair is accepted by an oxygen molecule, an additional $2H^+$ is removed from the matrix, convering $1/2C_2$ to $H_2O$. The electrons passed to cytochrome *b* are ready to enter another Q cycle.

In cycling from the Q to the $QH_2$ form each coenzyme Q molecule thus picks up one electron from an iron-sulfur protein and one from cytochrome *b*. In cycling back from $QH_2$ to Q, each coenzyme Q passes one electron to cytochrome *c and returns one to cytochrome b*. The Q cycle and the movements of coenzyme Q shuttling $H^+$ ions across the membrane from the matrix to the outside are considered to rest on the relatively small size of coenzyme Q and its solubility in the hydrophobic membrane interior.

### *Synthesis of ATP*

The hypothesis goes on to explain how the $H^+$ gradient established by electron transport powers ATP synthesis. The ATPase complex is considered to consist of two parts, $F_1$, located on the membrane surface facing the matrix, and $F_0$, located within the membrane, $F_1$ contains the site catalysing ATP synthesis or hydrolysis, $F_0$ is proposed to contain a channel that allows H* ions to approach the active site of the ATPase enzyme from the opposite side of the membrane. The membrane is otherwise considered to be impermeable to H+ ions.

The $H^+$ gradient established by electron transport results in net movement of $H^+$ ions from the intermembrane compartment into the $F_0$, channel, through which they approach the active site of the enzyme. On the matrix side of the membrane, the active site of the enzyme binds ADPOH and POH from the medium, converting them to the ionised forms $ADPO^-$ and $PO^-$. The $2H^+$ produced by this conversion are released into the matrix. The protons in the membrane $F_0$ channel now interact at the active site of the enzyme, removing one oxygen from $PO^-$ to form water, which diffuses back to the opposite side of the membrane. Removal of the oxygen causes conversion of ADPO and $PO^-$ to ADPOP, that is, to ATP.

$$ADPO^- + PO^- + 2H \rightarrow ADPOP + H_2O$$

Since $2H^+$ are released into the matrix by the conversion of ADPOH and POH to $ADPO^-$ and $PO^-$ and $2H^+$ are removed from the opposite side by combination with oxygen to form $H_2O$, the overall effect is the equivalent of moving $H^+$ from the side of high $H^+$ concentration in the intermembrane compartment to the side of low $H^+$ concentration in the matrix, thus relieving the "gradient. As a result, the total reaction tends to run in the direction of ATP synthesis as long as $H^+$ is available in excess in the intermembrane compartment.

### Experimental Support for the Mitchell Hypothesis

Mitchell's chemiosmotic hypothesis, at first largely rejected by the scientific community, has met with such complete experimental support that there now seems littie doubt that its two basic assumptions are correct : (1) Electron transport creates a gradient of $H^+$ ions, and (2) the $H^+$ gradient drives ATP synthesis. The hypothesis and the new insights into bioenergetic that is provides have revolutionised scientific thinking in oxidative phosphorylation, photosynthesis, and active transport. Only in some details does Mitchell's hypothesis, for which he received the Nobel Prize in 1978, remain open to question.

Tho olootron trancport creates an $H^+$ gradient is clearly supported by the available evidence. The Racker laboratory showed that closed vesicles can be isolated from mitochondria in either right-side-out or inside-out form. The sidedness of the vesicles can be easily identified by noting whether the ATPase lollipops are carried on the inside or outside surfaces of the vesical membranes.

If the vesicles are right side out, with the lollipops on the *inside* surface (analogous to their position in mitochondria, in which they extend into the matrix), they expel $H^+$ ions to the *outside* during electron transport. If inside out, with the ATPase lollipops on the outside surface, the vesicles concentrate $H^+$ ions on the *inside* during electron transport. Thus $H^+$ ions are actually expelled across mithochondrial membranes to the side opposite the ATPase enzyme as electron transport takes place, as proposed in the model.

Creation and maintenance of the $H^+$ gradient requires that the inner mitochondrial membranes be impermeable to $H^+$. Experimental support for this part of the hypothesis has come from Mitchell's laboratory. Mitchell and J. Moyle set up a pH gradient across mitochondrial membranes and rioted the rate at which $H^+$ could diffuse across the membranes to neutralise the gradient in the absence of electron transport and ATP synthesis. The rate of movement noted was so low that the inner mitochondrial membranes can indeed be regarded as essentially impermeable to $H^+$ ions.

Support for the second basic proposition of the hypothesis, that the $H^+$ gradient established by electron transport, drives ATP synthesis, comes from a number of sources. The first is the simple observation that closed, intact vesicles are necessary for ATP synthesis to accur in inner membrane preparations isolated from mitochondria. If the membranes are broken or made leaky, thus destroying any possibility of an $H^+$ gradient, ATP synthesis stops.

The second comes from observations of the effect of *uncouplers of* idative phosphorylation. These substances, when added to tochondria or mitochondrial vesicles, separate electron insport from ATP synthesis. Under the action of the coupler, electron transport runs continuously with no ationship to ATP synthesis. With few exceptions, the couplers, such as *DNP* (*2, 4-dlmitrophenol*), are lipid-soluble agents that cause membranes to be leaky to H+ ions. ,struction of the $H^+$ gradient removes the driving force for FP synthesis and stops phosphorylation, in accordance with e Mitchell hypothesis.

More direct evidence comes from experiments in which an imposed pH gradient was shown to cause ATP synthesis. The best of these was carried out in the laboratory of A. T. Jagenlorf, who used chloroplasts instead of mitochondria for several reasons. Chloroplasts contain an electron transport chain similar to the mitochondrial system, and ATP synthesis in :hloroplasts is coupled to electron transport in exactly the same way as in the mitochondria. The primary difference in ATP synthesis in the two organelles is in the source of high-energy electrons passing through the electron transport system. In mitochondria, these electrons come from oxidised fuel substances ; in chloroplasts, their energy is derived from absorbed light. By placing isolated chloroplasts in darkness, Jagendorf could eliminate electron transport as a source of energy for ATP synthesis.

He then created a surplus of $H^+$ ions inside the chloroplasts by placing them in a medium containing an acid that could penetrate inside. The chloroplasts were then transferred to a second medium at lower $H^+$ concentration. As the $H^+$ ions inside moved outside in response to the gradient, ATP was synthesized. Since electron transport was eliminated as an energy source, the ATP synthesis could be ascribed directly to the artificially created pH gradient. Similar experiments were subsequently carried out by Mitchell and his coworkers with mitochondria.

Another light-driven system has also provided an elegant demonstration that an $H^+$ gradient can provide the energy required for ATP synthesis. W. Stoeckenius discovered that the plasma membrane of a bacterium, *Halobacterium halobrium,* contains a pigment that resembles the visual pigments found in the rattn sof animal eyes. The pigment, called *bacteriorhodopsin,* can change rapidly into either of two forms, one that absorbs light at a wavelength of 570 nm ("*purple*") and one at 412 nm ("*bleached*"). At this pigment changes from the purple to the bleached form, it loses an $H^+$ ion, and in the opposite change, it picks up an $H^+$ ion. Racker and,

Stoeckenius found that purified rhodopsin, is combined with lipid bilayers in the form of closed vesicles, could take up $H^+$ ions from the medium and concentrate them inside the vesicles when exposed to light. If the components of the ATPase complex were then added hydrophobic membrane proteins plus stalk and $F_1$ proteins, the vesicles were able to drive the synthesis of ATP when illuminated! This simple system, containing nothing more than a closed membrane, a molecule establishing an $H^+$ gradient, and the ATPase enzyme, provides conclusive evidence that $H^+$ gradients can provide the energy required for ATP synthesis.

Experiments with ATPase-driven active transport systems also clearly support the idea that ion gradients can serve as energy sources for ATP synthesis. Normally the ATPasedriven pumps, such as the $Na^+$, $K^+$-ATPase system of the plasma membrane, break down ATP and use the energy released to establish an ion gradient.

However, the pumps can be made to run in reverse: Imposing an artificially high ion gradient can cause the pump's enzymatic system to run backwards and synthesize ATP from ADP and inorganic phosphate. These experiments directly show that the energy of an ion gradient can be used to drive ATP synthesis.

Taken together, the various experiments testing the iemiosmotic hypothesis establish beyond any reasonable doubt iat in mitochondria electron transport sets up an $H^+$ gradient, nd that the gradient in turn provides the energy source for iitochondrial ATP synthesis.

**Problems with Details of the Mitchell Hypothesis**

While the major and basic tenets of the chemiosmotic hypothesis are clearly supported by experiment, certain details of the model remain controversial. Most prominent among these is the Q cycle mechanism for expelling $H^+$ ions across the mitochondrial membranes. As of yet no evidence supports the unusual arrangement and sequence of electron carriers necessary for this mechanism.

Other difficulties stem from Mitchell's assumption that $6H^+$ will be expelled across the membrane for each electron pair traversing the entire chain from NAD to oxygen. Actual measurement indicates that $6H^+$ may be expelled for each electron running through the carrier sequence, or $12H^+$ per electron pair.

In another sense, however, finding a higher ratio than predicted supports other, more basic tenets of Mitchell's model. One persistent trouble with the model has been that, with only $6H^+$ expelled for each electron pair traversing the chain, most calculations (except for Mitchell's) have shown that the $H^+$ gradient created would not be quite high enough to power synthesis of the three ATP molecules formed when a pair of electrons travels the entire route from NAD to oxygen.

Increasing the level to 4-5 $H^+$ expelled per electron, or 8-10 $H^+$ per electron pair, would build a gradient high enough to satisfy all calculations of the energy required. These difficulties are relatively minor ones, however, and do not detract from the fact that in his brilliant hypothesis, Mitchell has successfully deduced the biochemical mechanims underlying oxidative phosphorylation.

## The Conformational Hypothesis

The conformational hypothesis of P.D. Boyer is of special interest because it provides a possible explanation for the details of ATP synthesis in the chemiosmotic hypothesis. Boyer's idea is based on the model currently accepted for force generation in muscle. In muscle, *myosin,* a molecule with ATPase activity changes its conformation or folding pattern as it binds and hydrolyses ATP. The conformational change imposes a strain on parts of the myosin, which is relieved by physical movement of a part of the molecule. The movement is multiplied by the millions of myosin molecules involved and is translated into movement of the entire muscle. In the muscle system, therefore, ATP breakdown is considered to cause strain and movement.

In Boyer's application of this mechanism to oxidative phosphorylation, this process operates in reverse : Electron transport and the resultant $H^+$ gradient lead to a conformational change in the mitochondrial ATPase, producing a strain in the structure of the molecule. This strain is relieved by the synthesis of ATP.

Conformational changes may be important in ATP synthesis as Boyer suggests. They may also be important in the mechanisms setting up the pH gradient through electron transport. For example, the expulsion of $H^+$ across cristae membranes could be due to conformational changes in the carrier proteins. In this case, a cycle of reduction and oxidation of a carrier molecule (or complex of carriers) could lead to a conformational change exposing an $H^+$-bearing chemical group such as a carboxyl (—COON) at one surface of the membrane. In the new position, the group might dissociate more easily, leading to release of $H^+$ at this side of the membrane.

## Location of the Components of the Electron Transport and ATP-Synthesizing Systems

The molecules carrying out electron transport and ATP synthesis were identified with the cristae membranes by experiments following the distribution of biochemical activity after disruption of mitochondria. At low concentration, a detergent such as digitonin breaks the outer mitochondrial membrane without damaging the inner membrane. The membranes can then be separately purified by centrifugation.

After removal of the outer membranes, the capacity for electron transport remains with the cristae membranes as long as intact, closed vesicles are retained. All of the various components of electron transport and ATP synthesis are found to be linked to the cristae membranes, rather than suspended in solution in the matrix.

Application of an extensive battery of newer techniques has begun to reveal the locations and arrangement of these components within the cristae membranes. The most productive of these methods utilise vesicles isolated from inner mitochondrial membranes oriented in right-side-out or insideout fashion.

These vesicles have been tested by three different probes : (1) antibodies developed against various components of the electron transport and ATP-synthesizing systems ; (2) reagents that chemically modify or attach radioactive labels to the parts of proteins exposed on the inner or outer membrane surfaces, and (3) nonpenetrating electron acceptors or donors that react specifically with individual carriers of the electron transport system. None of the probes used can penetrate across the mitochondrial membranes. Basically, the probes test which of the two surfaces of the inner mitochondrial membranes, either the side facing the matrix (the *M face*) or the side facing the membrane space (the IM *face*), carry exposed portions of the molecules "*recognised*" by the probes.

The work with cytochrome a serves as a good example of the approach and the conclusions drawn. Antibodies made against cytochrome a interact with the IM face, but not the M face, of vesicles derived from mitochondrial cristae. Polylysine a reagent capable of linking to the parts of proteins exposed at membrane surfaces, binds to cytochrome a only is applied to vesicles with the IM face exposed. From. this information, cytochrome a is assigned an asymmetric location in the cristae membranes, in the membrane bilayer half facing the intermembrane space.

Combination of the results from several laboratories, notably those of Racker and R.A. Capaldi reveals that most of the molecules functioning in electron transport and ATP synthesis share cytochrome as one sided orientation and face only one of the two inner mitochondrial membrane surfaces. The single exception is succinic acid dehydrogenase, the enzyme carrying FAD as prosthetic group, which apparently spans the membrane and has parts exposed on both surfaces.

The active site of the enzyme, however, reacts with succinic acid only on the M face and is evidently directed towards this side. Locating coenzyme Q, the only electron carrier not linked to a protein, has proved difficult. Most experiments indicate that Q is buried entirely within the membrane. Chance, for example, observed that the quinone ring of coenzyme Q could react only with reageats possessing hydrophobic groups that can penetrate into the membrane interior.

Comparing the absolute ratios obtained when components of the electron transport system are extracted and purified reveals that the individual carriers occur in cristae membranes in the ratios 1 FAD : 1 FMN : 7-10 coenzyme Q : 2 cytochrome *b* : I cytochrome $c_1$ : 2 cytochrome *c* : 2 cytochrome *x* : 2 cytochromc $a_3$. Information from this and other approaches also indicates that there is one ATPase complex for each electron transport system. These ratios suggests that the carriers are coupled into units that contain the individual carriers in fixed proportions, together with one ATP synthesizing complex per unit.

If this is the case, it is interesting to note that the proportions of the carriers are apparently unrelated to whether the individual carriers transport electrons singly or in pairs. That is, single and pair electron carriers are not necessarily present in a 2 : 1 ratio, as might be expected.

**Location of Other Components of Oxidative Metabolism in Mitochondria**

All but one of the enzymes and substrates of the citric acid cycle, and the enzyme complex carrying out pyruvic acid oxidation, are released from the interior of mitochondria immediately on rupture of the inner mitochondrial membranes. Removal of the outer membrane alone, however, does not release these factors. Therefore, these enzymes and substrates are considered to be in solution in the matrix in the compartment enclosed by the inner membrane. The single exception is succinic acid dehydrogenase, which, with its prosthetic group (FAD), forms a part of the electron transport system and is tightly bound to the cristae membranes as an integral membrane protein.

Enzymatic activity of the outer mitochondrial membrane is relatively limited. One enzyme, associated with the oxidation of amino acids, an amine oxidase, is so characteristic of the outer mitochondrial membrane that it is used as a marker to identify this membrane fraction in isolated preparations. Other enzymes have also been identified within the outer membranes, including the complex oxidising pyruvic acid.

Given the extreme complexity and variety of the enzyme systems present in mitochondria, which, in addition to oxidative enzymes, also include full facilities for DNA and protein synthesis, these organelles are probably the most biochemically complicated and capable structures inside animal cells. In plant cells they are rivaled only by chloroplasts, which, as the next chapter will show, are equally diverse in their biochemical activity.

## Cellular Oxidation of Fats and Proteins

Both fats and proteins, as noted at the beginning of this chapter, can serve as cellular fuels. As a preliminary to their cellular oxidation, these comparatively large molecules are first broken into smaller units : Fats into glycerol and fatty acids and proteins into amino acids. The smaller units are then oxidised in reactions that take place primarily inside mitochondria. The individual amino acids derived from hydrolysis of proteins are oxidised in a variety of pathways.

These pathways, depending on the particular amino acid, eventually yield products that enter the main line of carbohydrate oxidation as either pyruvic acid, acetyl coenzyme A, or intermediates of the Krebs cycle. As a part of the pathways leading to these entry points, the amino acids are deaminated (the $—NH_2$ group is removed).

Products of both fat and protein breakdown thus enter the central carbohydrate pathway to be oxidised to $CO_2$ and $H_2O$ primarily in mitochondria summarises these routes. Note in this figure the central role played by coenzyme A as a carrier funneling products of different pathways into the Kerbs cycle.

The end product of all of these reactions that is significant to cellular activity is ATP. Additional oxidative pathways of importance are summarised in the supplements to this chapter. Supplement describes the *pentose phosphate pathway*, an alternate route for carbohydrate oxidation that also serves as a source of five-carbon sugars. Additional pathways for the oxidation of fats and amino acids, taking place in microbodies (peroxisomes and glyoxisomes), are discussed in Supplement.

## Integration of Mitochondrial Activity with the Cell Environment

### *Mitochondrial Transport*

The reaction pathways of pyruvic acid oxidation, fatty acid Dzidation, and the Krebs cycle are located inside the mito,^,hondrial matrix. How are the substrates for these reactions and the ADP-ATP couple transported in and out of mitochondria ? The outer mitochondrial membrane is freely permeable to most organic and inorganic molecules with molecular weights up to about 5000. Both mitochondrial membranes freely admit $H_2O$, $O_2$, and $CO_2$.

However, the inner mitochondrial membrane surrounding the matrix is almost completely impermeable to most hydrophilic substances, except for a group of molecules that are transported by specific carrier systems.

Several of these systems have been identified reviewed in LaNoue and Schoolwerth. One, called the *adenine nucleotide carrier,* exchanges ATP for ADP in either direction across the inner mitochondrial membrane. Another carrier mechanism, already mentioned, shuttles electrons carried by reduced NAD outside to NAD inside the mitochondrion. Other specific carriers transport intermediates of the various oxidative pathways located in the matrix, including inorganic phosphate, fatty acid derivatives, pyruvic acid, various amino acids, and acids of the Kerbs cycle.

Various lines of evidence indicate that the inner membrane carriers are proteins. They are highly specific for the molecules or molecular groups carried and exhibit saturation at high concentrations of the transported substances. Some of the carrier proteins, including the adenine nucleotide carrier exchanging ATP for ADP across the inner membrane have been isolated and partially purified. Addition of the nucleotide carrier protein to artificial phospholipid bilayers transfers limited ATP-ADP transport capability to the bilayers.

Most of the transport systems of the inner membrane are active, and thus require energy for their functions. In mitochondria, the carriers are evidently driven by the $H^+$ gradient set up across the inner membrane by electron transport. The electron transport system activity expels $H^+$ ions into the intermembrane compartment, setting up the $H^+$ gradient. Diffusion of the $H^+$ ions back into the matrix is tightly coupled to the inward transport of the substances admitted by the inner membrane carriers. Transport of these substances thus occurs by the H+-linked cotransport pathway.

***Regulation of Mitochondrial Oxidation***

The rate of mitochondrial oxidation is regulated by a system of controls within the cell. Of these, the most important is based on the concentration of ADP in the medium surrounding the mitochondrion. As ADP concentration increases, phosphorylation of ADP to ATP begins inside mitochondria. This phosphorylation is closely coupled to electron transport; as phosphorylation takes place, electrons flow without restriction from NAD to oxygen, and the NAD end of the transport chain becomes relatively oxidised. NAD, once oxidised, is free to recycle as an electron acceptor in reactions oxidising the substrates of the Krebs cycle. This system closely links the rate of ox dation in the Krebs cycle to the concentration of ADP in the cell.

If cellular activity is limited. ATP concentration will remain high. and ADP will become unavailable to the ATPsynthesizing enzyme complex in cristae membranes. As a result, the ATP-synthesizing reactions and electron transport will stop. According to the Mitchell hypothesis, stoppage of electron transport is due to a buildup in the $H^+$ gradient that is unrelieved by ATP synthesis. The gradient gradually builds until it reaches levels high enough to oppose further expulsion of $H^+$ across the cristae membranes by the electron transport system. In consequence, electron transport and, in turn, oxidation in the Krebs cycle stop.

**Table 4.3: Transport Carriers of the inner Mitochondrial Membrane.**

| *Carrier* | *Activity* |
|---|---|
| Adenire nucleotide | Exchanges A UP for ADP |
| Phosphate | Transports $HPO_4^2$ |
| Pyruvic acid | Transports pyruvic acid |
| Glutamic acid | Transports glutamic acid |
| Tricalboxylic acid | Exchanges citric, isocitric, malic, and phosphoenolpyruvic acids |
| Dicarboxylic acid | Exchanges malic, malonic, and succinic acids |
| α-ketoglutaric acid | Exchanges a-ketoglutaric, malic, succinic, and oxaloacetic acids |
| Glutamine | Transports glutamine |
| Carnitine | Transports fatty acid derivatives |

Control of glycolysis may also be linked to mitocbondriat activity through the oxidation-reduction cycle of extramitochondrial NAD. Continuance of glycolysis depends on re oxidation of the NAD reduced in the sequence. Through the NAD shuttle systems, extramitochondrial reduced NAD may transfer its electrons to NAD or FAD inside the mitochondrion to become reoxidised and cycle back to glycolysis. Passage of electrons through the shuttle systems depends on the availability of oxidised NAD or FAD inside the mitochondrion.

This in turn depends on the rate of electron transport and ADP concentration. As cells carry out their major activities, including growth, movement, and response to their environment, ATP is hydrolysed to ADP. The increase in ADP concentration causes an increase in the rate of oxidation in the mitochondrion, and the concentration of ATP is restored. All the various activities, from glycolysis through synthesis, thus exists as a delicately balanced system, in which the oxidative mechanisms are finely -tuned to the cell's energy requirements.

## ORIGIN OF MITOCHONDRIA

Our understanding of the process where by mitochondria are produced, is still very incomplete. *Lehninger* classified the various theories of possible routes of mitochondria) genesis into three main groups

(1) Formation from other membranous structures in the cell,

(2) Growth and division of pre-existing mitochondria.

(3) *De novo* synthesis from submicroscopic precursors.

### *Formation from other membranous structures in the cell*

The formation of mitochondria by "pinching off" or budding from pre-existing cell structures has been suggested for a range of cell membranes including those of the plasmalemma, endoplasmic reticulum, nuclear envelope, and Golgi complex. But the support for such evidence, in the absence of supporting biochemical data, can not be wholly conclusive.

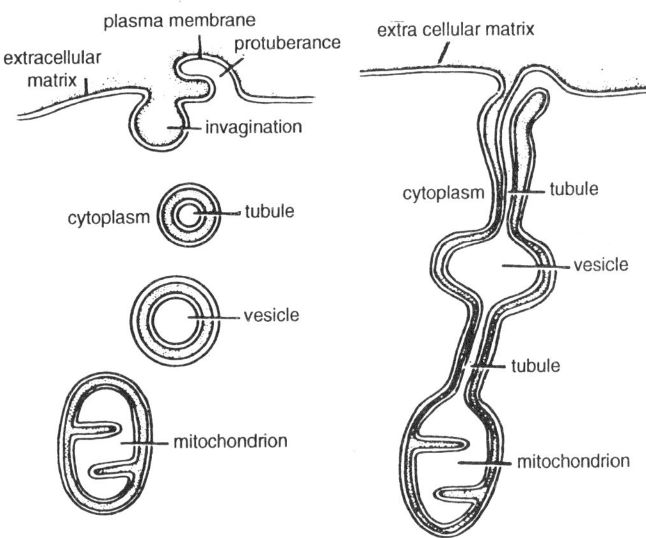


*Figure 4.10: Origin of mitochondria from plasma membrane or endoplasmic reticulum.*

Part of the problem undoubtedly lies in our fragmentary knowledge of the structure and composition of and differences between cell membranes in general. Indeed the similarities discussed in the literature between the mitochondrial membrane and the endoplasmic reticulum, could lend weight to the idea that probably mitochondria are formed when cytoplasm pushes into a cavity surrounded by an internal membrane, which then 'pinches off' and separates from the continuous system.

***Growth and division of pre-existing mitochondria***

Electron microscopic evidence mitochondrial division by fission, although plentiful is difficult to assess the danger of producing artifacts is very real because of the harsh chemical and physical agents brought to bear on the test material during processing. Interpretation is not made easier by thetability of mitochondria to undergo extreme changes in shape *in vivo* which may or may not be associated with mitochondrial fission.

There are numerous reports of mitochondria connected to each other by narrow bridges of membrane, especially in rapidiy me'opolising tissue, and it is thought that such figures may . represent mitochondria in an early stage of fission. By observing serial sections of rat liver *Stempak* (1967) was able to show that "dumb bell-shaped" mitochondria can be the sections of cupshaped bodies. Such bodies have also been observed in rapidly growing tissues of fern and may represent the begining stages of division.

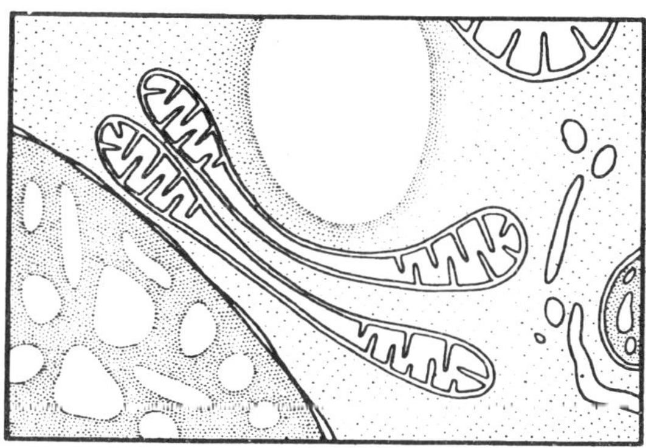

*Figure 4.11: Division of mitochondria by elongation.*

An early stage in mitochondria) division may involve the separation of mitachondrial contents into two or more compartments. The presence of mitochondria with internal "partitions" has been formed in several cell types although the possibility that they are manifestations of mitochondrial fusion can not easily be ruled out. *Lafontaine* and *Allard* have presented electron micrographs of rat liver mitochondria which exhibit what appear to be partitions dividing the inner membrane complex into two masses, the whole being surrounded by continuous outer membrane.

*Tandler et. al.* has demonstrated the partitions of mitochondria in liver which was recovering from riboflavin deficiency.

***De novo synthesis***

The possibility of *de-novo* synthesis of mitochondria arose with experiments in the early part of the century, when mitochondria containing larvae were seen to develop from sea urchin egg cytoplasm which had apparently been freed of mitochondria by centrifugation Using the greater resolving power of the electron microscope, it was later shown, that mitochondria could not be dislodged by centrifugation of the egg.

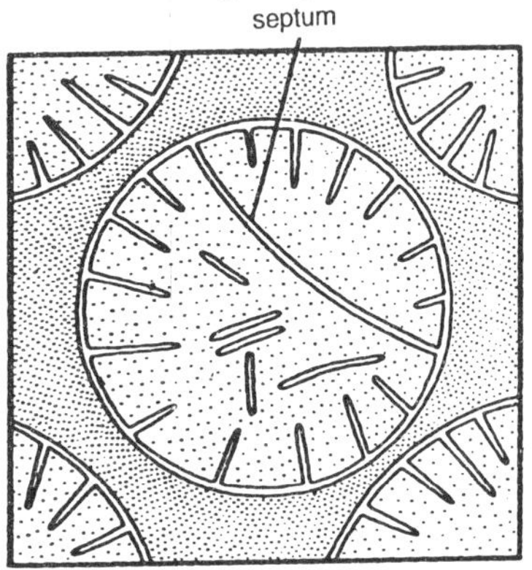


*Figure 4.12: Division of pre-existing mitochondria by the formation of septa.*

In the earlier experiments, mitochondria had probably been present in the "*centripetal end*" of the egg cell after all, and these mitochondria could have served as precursors in subsequent mitochondrial production. In the above description, a number of views were described, regarding the generation of mitochondria in the cytoplasm of different types of cells. But it is perhaps unwise to group the evidence to suit one or other of a limited number of clear cut methods by which mitochondria could replicate.

The actual situation is probably complex, and it may well be that different methods of replication take place in different tissues, and at different stages in development. One could imagine the early mitochondria being formed from membrane structures in the developing embryo concentration of mitochondria around the nuclear membrane have been noted in embryonic tissues from several phyla and formation of mitochondria from this membrane could involve the transfer of nuclear genetic information essential for subsequent mitochondrial growth and multiplication by division.

The multiplication of mitochondria could then proceed by the incorporation of large pre-fabricated molecules and association of molecules, with division by fission, when the mitochondria reached a critical stage.

**Prokaryotic Origin of Mitochondria**

The fact that mitochondria can grow, divide and are capable Mutations support a long-held view that mitochondria originated with their host. Bacteria would have originated the mitochondria and blue green algae, the chloroplasts.

There are numerous homologies between mitochondria and bacteria. In bacteria the electron transport system is localised in the plasma membrane which can be compared with the inner membrane of mitochondria. Some bacteria even have membranous projections extending from the plasma membrane which are comparable to mitochondrial crests since both contain the respiratory chain.

The inner membrance and matrix, it has been postulated may represent the original symbiont which may become enclosed within a membrane of cellular origin (ER). Further mitochondrial DNA is circular ; it replicates and divides like that of bacteria. The ribosomes are also found which are, however, smaller than those of the bacteria. In mitochondria and bacteria, the protein synthesis is inhibited by chloramphenicol.

From these similarities, one can easily conceive of mitochondria as being evolved from an ancient prokaryote possessing all the attributes of an independent, probably aerobic organism. However, with the adaptation over a long period it became an essential and dependent symbiont, lost same of its identity to the cell, and conversely, the host cell lost some of its functions, deriving it now from the endosymbiont or mitochondrion. As a results, both became obligatory symbionts to each other.

This spmbiont hypothesis for the origin of mitochondria and plastids has achieved wide popularity, but all biologists do not necessarily accept it. *Raff* and *Mahler* concluded that "while the symbiotic theory may be esthetically pleasing, it is not compelling." They presented a lot of evidences and proposed that mitochondria arose by inward blebbing from plasma membrane, by the acquisition some how of an outer membrane, and by the *additional* acquisition of a DNA genophore from the DNA of the protoeukaryote in which the evolution of mitochondrion occurred. *Borst* proposed an *episone theory* and supposed that the DNA of mitochondrion left the '*nuclear*' DNA by a sort of amplification to become mapped within a membrane containing the respiratory chain.

# 5

# Golgi Body

In 1898 by means of a silver staining method, *Golgi* discovered a reticular structure in the cytoplasm. The name "Golgi apparatus", generally given to this structure, is confusing because it suggests a definite relationship with the physiologic processes of the cell. Today it seems more appropriate to use the name "*Golgi substance*" or "*Golgi complex*", to refer to this material that has special staining properties. Because its refractive index to similar to that of the matrix, the Golgi complex is difficult to observe in living cells. The use of the electron microscope has provided a distinct image of this component, and its submicroscopic structure has been revealed.

For years the Golgi complex was thought to be an artifact of various fixation and staining procedures. In other words, many scientists believed that the structure observed during numerous microscopy procedures and termed the Golgi did not actually exist in the living cell. Guilliermond, *Parat Walker* and *Allen,* doubts arose with regard to the existence of Golgi complex.

The different lines of objections against the existence of Golgi complex are as follows:

(*a*) Nothing corresponding to the Golgi elements exists in the living cells, what appears Golgi bodies in the processed cells are myeline figures, produced by the action of cytological reagents or lipids, hove. ever, believe in the existence of Golgi bodies in the germ cells but no such things in the somatic cells.

(*b*) The Golgi apparatus of the fixed material is produced by the deformation and metallic impregnation of several unrelated cytoplasmic bodies.

(*c*) The dictyosome or acroblasts of male germ cells iomologised with the Golgi apparatus of nerve cells. *Gatenby*, *Bowen* and *Hirschler*, *Parat* believed hat they are nothing but the modified mitochondria.

(*d*) The Golgi apparatus is a stacks of the smooth endoplasmic reticulum.

## ANSWER TO THE OBJECTIONS

(*a*) In a number of cells those bodies have been studied in *vivo* or in *vitro* with phase contrast microscopy.

(*b*) The Golgi bodies have been separated by centrifugal fractionation of homogenised cell. The objection of artifact is discarded on the basis that artifact can not show the pattern of the same type as shown by these bodies.

(*c*) Their ultra structure has been recently studied with the electron microscope by *Sjostrand* and *flanzon* and *Dalton* and *Felix*.

(*d*) Dictyosomes of male germ cells which are impregnated with silver and osmium, have same type of chemical composition as the Golgi apparatus of somatic cells.

(*e*) According to *Srivastava* dictyosomes in male germ cells, are favourable materials for the study of Golgi bodies even under the light microscope.

(*f*) The idea that the dictyosomes are modified mitochondria has been discarded on the basis of biochemical tests.

(*i*) The dictyosomes of male germ cells contain appreciable quantity of polysaccharides, it is also present in the somatic Golgi bodies but not found in mitochondria. Similarly alkaline phosphatase is associated with Golgi bodies, but not found in the mitochondria.

(*ii*) *Tiwari* and *Broune* have found that the distribution of adenosine triphosphatase is localised in mitochondria but not in Golgi bodies. In the same way succinic dehydrogenase and cytochrome oxidase are noted from the mitochondria, but not in the Golgi bodies.

(*g*) The statement that the Golgi bodies of neuron are nothing but endoplasmic reticulum, is discarded on the basis of electron microscopic structure and the width of Golgi membrane is different from that of the endoplasmic reticulum.

In concluding the controversy of the Golgi apparatus, it should be pointed out that *Baker* and this students were the last to accept the reality of the Golgi apparatus when *Baker* wrote "The author accepts, after long hesitation, the view that the Golgi apparatus in the neurons of vertebrates corresponds with the organelle of the same name, in other cells".

## TERMINOLOGY OF GOLGI COMPLEX

*Holmgren* referred to the Golgi complex as *trophospongium* (*Cajal* referred it as the *Golgi-Holmgren canals*). *Baker* used the term *lipochondria* because of the presumed lipid content. The term *Dalton Complex* was given after the name of its observer *Dalton* in 1952. *Sjostrand* proposed the term cytomembranes for the Golgi system.

*Sosa* has suggested the following nomenclature for the Golgi Complex.

(*a*) *Golgiokinesis*. Division of the Golgi apparatus during nuclear division.

(*b*) *Golgiosomes*. Corpuscles produced by the Golgiogenesis are called as Golgiosomes which are described as Golgi material in invertebrates.

(*c*) *Golgiolysis*. Process of dissolution of the Golgi apparatus.

(*d*) *Golgiorrhexis*. Fragmentation of the Golgi apparatus.

(*e*) *Golgiogenesis*. Formation and differentiation of the Golgi body during embryology.

(*f*) *Golgi-cytoarchitecture*. Study of structure of cell in relation to Golgi apparatus.

## OCCURRENCE

The Golgi complex occurs in all cells except the prokaryotic -ells (viz., mycoplasma, bacteria and blue green algae) and °ukaryotic cells of certain fungi, sperm cells of bryophytes and pteridophytes, cells of mature sieve tubes of plants and mature sperm and red blood cells of animals. Their number per plant cell can vary from several hundred as in tissues of corn root nd algal rhizoids (i.e., more than 25,000 in algal rhizoids, *Sievers*, 1965), to a single organelle in some algae. In animal ells, there usually occurs a single Golgi complex, however, eveloping oocytes of chordates may contain many Golgi bodies.

## MORPHOLOGY OF GOLGI COMPLEX

The morphology of the Golgi complex varies from cell to cell depending upon the type of cell in which they are found. Two forms of Golgi complex have been observed.

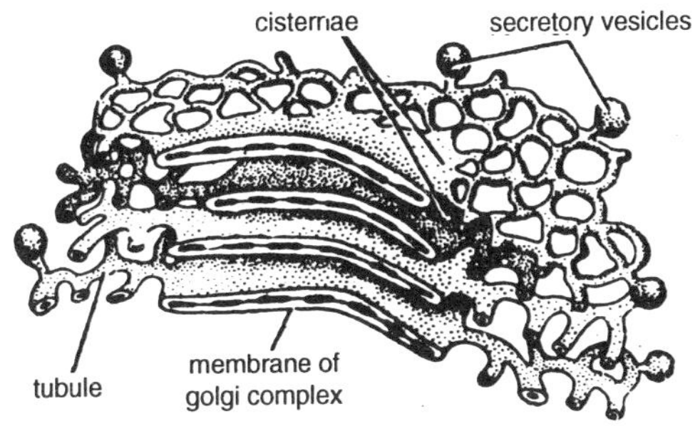


*Figure 5.1: Three dimensional view of Golgi complex.*

**Localised form**

In polarised cells of vertebrates (which have base and apex), Golgi complex occurs singly and occupies a fixed position. It lies between the nucleus and secretory pole. This can be best seen in thyroid cells, in exocrine cells of pancreas and the mucous cells of intestine.

**Diffused form**

In some specialised cells of vertebrates (nerve cells and liver cells), in most plant cells and in the cells of invertebrates several units of Golgi complex are found scattered along with the elements of endoplasmic reticulum. Each unit is called a *dictyosome*. In liver cells there occur as many as 50 dictyosomes per cell and in certain plants cells their number may reach upto hundreds.

## SHAPE

The shape of the Golgi complex is quite variable in different somatic cell types of animals. Even in the same cell there are variations in different functional stages. The shape is, however, constant with each cell type. It varies in form from a compact mass to a disperse filamentous network.

## NUMBER

The number of Golgi stacks per cell varies enormously, depending on the cell type-from as few as one to the hundreds. There is a single large in some cells while in the case of *Paramoeba* there are two. In *Stereomyxa* (a species of *Amoeba*) there ate many Golgi complexes. Nerve cells, liver cells and most plant cells also have multiple Golgi complexes, there being about 50 in liver cells. The Golgi complex can even account for a large fraction of the cell volume in some specialised cells. One example is the goblet cell of the intestinal epithelium, which secretes mucus into the gut; the glycoproteins in mucus are glycosylated principally in the Golgi complex.

**Size**

The size is likewise very variable. It is large in the nerve and gland cells and small in the muscle cells. In general the Golgi complex is well developed while the cell is in active state. When the cell grows old, the apparatus progresively diminishes in size and disappears.

**Position**

The position of the Golgi complex is relatively fixed for !h cell type. In the cells of octodermal origin, the Golgi nplex is polarised from the time of the embryonic state .weep the nucleus and the periphery. In the secretory exocrine cells that have in general a typical polarisation the ilgi complex is found between the nucleus and the secretory pole. In the endocrine glands the polarity of this organoid is variable, except in the thyroid, where it is oriented towards the itre of the follicle. In the younger cells and often in the older ones it lies most commonly at one side of the nucleus but certain cases may completely surround it. In ganglionic cells the mouse the position is perinuclear.

## DETAILED STRUCTURE OF GOLGI COMPLEX

*Dalton* and *Felix* described the Golgi complex in the Jidymis of the rat after taking the first electron microphs. The following description of the Golgi complex is a aposite one based on the work of several authors.

**Cisternae**

The cisternae or saccules are similar to the smooth surface ER, and appear in section as stacks of closely spaced membrane-delimited sacs. The number of saccules varies from about 4 to 8 in most animal and plant cell types. In *Euglena*, the number may go upto 20. The membrane of the saccule is roughly 60 to 70Å in thickness which encloses a cavity about 150Å wide whose edges are often dilated. According to most authors, there are two well defined faces of the cisternae, i.e., convex and the concave ; the later is generally referred to as e mature or forming or distal face and the convex side is assumed to be the immature or excit or proximal face, the cisternae lie in parallel array are separated from each other by a space of about 200

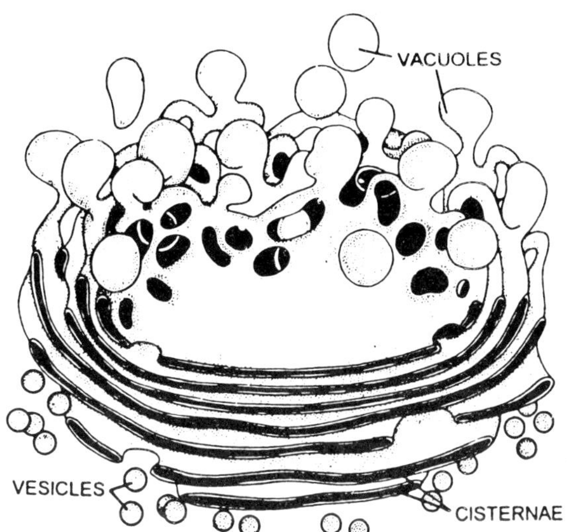


*Figure 5.2: Three dimensional veiw of Golgi complex.*

to 300Å. What holds them together is not yet known but in few cells a thin layer of electron opaque, sometimes dense material is seen between th saccules which at certain regions are more prominent to which *Amos* and *Grimstone* applied the term nodes. *Mollenhauer et. al.*, explored in some detail intercisternal elements and plaques in certain plant Golgi complex.

**Tubules**

From the peripheral area of cisternae arise a complex, anastomosing flat network cf tubules of 300 to 500Å diameter. *Clowes* and *Juniper* have compared this tubular network to disc of lace.

**Vesicles**

The vesicles are small droplet-like sacs which remain attached to tubules at the periphery of the cisternae. They are of following two types

***Smooth vesicle***

The smooth vesicles are of 20 to 8m0μ diameter. They contain secretory material (so often are called secretory vesicles) and are budded off from the cisternal tubules within the net. Often more than one tubule connect to, and presumably fill, a single forming vesicles.

***Coated vesicles***

The coated vesicles are spherical protuberances, about 50 m$\mu$ in diameter and with a rouge surface.

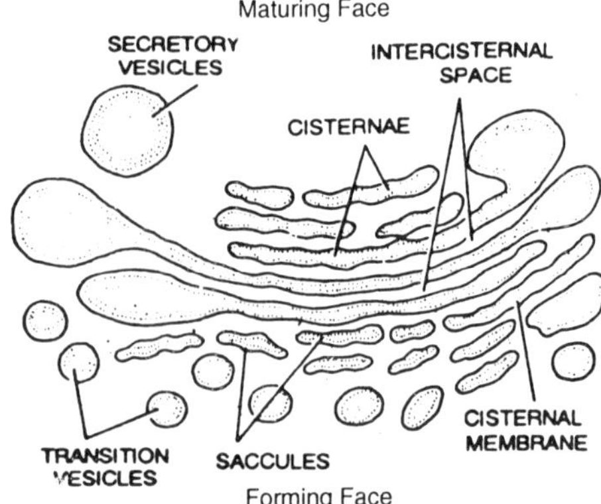


*Figure 5.3: Ultrastructure of Golgi complex.*

They are found at the periphery of the organelle, usually at the end of single tubules, and are morphologically quite distinct from the secretory vesicles. Their function is unknown.

***Golgian vacuoles***

These are large rounded sacs present on the maturing face of Golgi. These are formed either by the expanded cisternae or by the fusion of secretory vesicles. The vacuoles are filled with some amorphous or granular substance.

## THE GOLGI COMPLEX IS STRUCTURALLY AND BIOCHEMICALLY POLARISED

The Golgi complex has two distinct faces: a *cis*, or *forming face* and a *trans*, or *maturing face*. The cis face is closely associated with a smooth transitional portion of ther ough ER. In secretory cells, the trans face is the face closest to the plasma membrance; here, the large secretory vesicles are found exclusively in association with the trans face of a Golgi stack, and the membrane of a forming secretory vesicle is often continuous with that of the trans face of the last ('*transmost*") cisterna.

In contrast, the small Golgi vesicles are localised more evenly along the stack. Proteins are commonly thought to enter a Golgi stack from the ER on the cis side and to exit far multiple destinations on the trans side; however, neither their exact path through the Golgi complex nor bow they travel from cisterna to cisterna along each stack are known.

The two faces of the Golgi complex are biochemically distinct. For example, a variation in the thickness of the Golgi membranes can be detected across the stack in certain cases, with those at the cis side being thinner (ER-like) and those at the trans side being thicker (plasma-membranelike). More striking are the results obtained when certain histochemical tests are used in conjunction with electron microscopy to localise particular proteins within the Golgi complex. Some of these tests reveal membrane-bound enzyme activities that show a distinct polarity in their localisation within the Golgi stack.

A particularly intriguing biochemical finding was the discovery that lysosomal enzymes, such as acid phosphatase, are concentrated with the trans-most cisterna of the Golgi stack and within some of the coated vesicles nearby. This suggests that specific vesicles leaving for lysosomes are assembled in this region.

Secretory proteins are found by histochemical methods in all of the stacked cisterna, even though the large secretory vesicles in which these products are concentrated and associated only with the trans-most Golgi cisterna.

## CHEMICAL COMPOSITION

Regarding the chemical composition of Golgi complex, it has been demonstrated that the following substances are present:

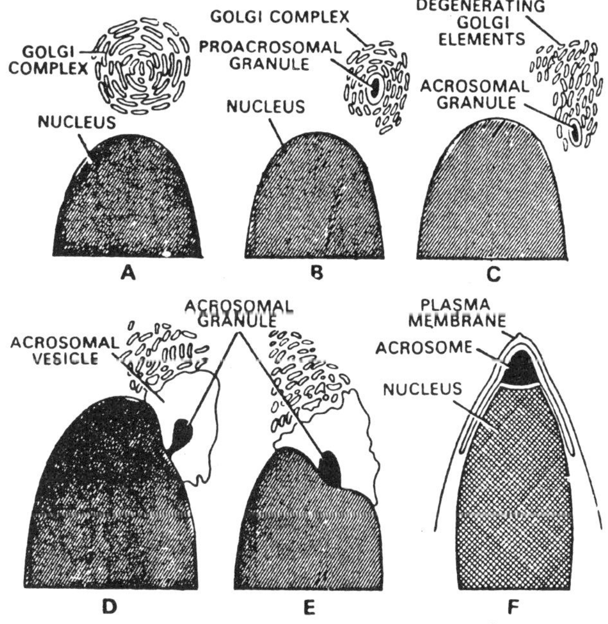


*Figure 5.4: Role of Golgi complex in the formation of acrosome.*

**Phospholipids**

Phospholipids composition of Golgi membranes is intermediate between those of endoplasmic membranes and plasma membranes.

**Proteins and Enzymes**

Golgi complex from different plant and animal cells show great variations in the protein and enzyme contents. Some of the enzymes are ADPase, ATPase, CTPase, NADPHcytochrome-C-reductase, glycosyl transferases, galactosyl transferase, thiamine pyrophosphate etc.

**Carbohydrates**

Both plant and animal cells have some common carbohydrate components, like glucosarine, galactose, glucose mannosa, and fructose. Plant Golgi lack sialic acid, but occurs in high concentration in rat liver. Some carbohydrate like xylase and arabinose are present in plant cell only.

**Vitamin C**

The function of vitamin C stored in the Golgi complex has been shown by *Tomitte.* According to him Golgi complex stores vitamin C and liberates it slowly into the cytoplasm it sufficient amount to prevent-oxidation of the cell products.

## FUNCTIONS OF GOLGI COMPLEX

**Formation of Acrosome during spermiogenesis**

During the maturation of sperm the Golgi complex plays a in the formation of acrosome. In early stages, the Golgi appears as a spherical body, comprising cisternae arranged in paralled stacks and numerous small vesicles. The later always pinched off from the cisternae. As development proceeds, the Golgi complex becomes irregular in shape and large vacuoles are formed by dilations of cisternal sacs.

In the centre of these large vacuole or vacuoles is/are present a dense granule, the *proacrosomal granule.* This granule which is derived from Golgi complex continues to grow within the vacuole by a process known as *accretion.* This vacuole and granule approaches the anterior pole of the nuclear membrane, constituting *acrosomal granule.* With the elongation of the spermatid, the *acrosomal vesicle* spreads over the nuclear surface and finally collapses with the nuclear membrane, forming the *cap material.* The acrosomal granule becomes the *acrosome* which lies at the apex of the nucleus and apparently comprises certain enzymes involved in the process of fertilisation.

**Synthesis and secretion of polysachharides**

Studies on goblet cell by autoradiography and electro microscopy have established the inter-relationship between protein synthesis, carbohydrate addition and sulphation. The goblet cells of the colon produce mucigen. This secretory naterial contains a large proportion of carbohydrate. The Golgi complex is found just above the nucleus. Towards the ree surface of the cell are gradully enlarging mucigen granules. Che proximal cisternae of the Golgi complex do not show any welling, but at some distance across the stack the distal isternae are quite suddenly converted into mucigen granules. The distal cisternae continually convert into mucigen granules very 2-4 minutes. New proximal cisternae are formed in compensation.

**Role in Secretion**

Golgi complex is considered to play some role in the secretory functions of a cell. But the question is this that they are secreting or synthesizing some substances themselves or they are simply a store house in which the secrtory products which are secreted some where else in the cell, is simply ored and concentrated.

From the studies of *Palade et. al. 1962* this secretory is now well defined and includes four steps in case of pancreatic acinar cells and they are

(*a*) The in-corporation of aminoacids into protein at the surface of rough endoplasmic reticulum.

(*b*) Transfer of these nascent secretory proteins into the cisternae of rough endoplasmic reticulum.

(*c*) The intracellular transport of these proteins to the Golgi complex.

(*d*) The migration of *zymogen granules* towards the apex of the cell where they discharged into lumens.

**Role of Golgi body in oogenesis**

*Srivastava* has given a brief review on the Golgi complex during oogenesis.

According to *Afzelius*, the Golgi complex of a seaurchin egg, as seen under electron microscope, consists of stacks of lamellae forming walls of flat pouches, which may occasionally be swollen. There is some indications of transverse divisions of these bodies. *Satelo* and *Solelo* and *Porter* have described the Golgi complex in rat-ovum as seen under the electron microscope and found juxtra nuclear localisation of this organelle in early oocytes. In the next stage, these resolve into fragments and in the third stage, these move towards the cortex. In all these cases, their structures remain to be of closely packed arrays of slender, double profiles (flattened sacs) and spherical vesicles. In the early oocytes the complex is compactly organised. In later stage, discrete bundles of profiles, surrounded by small vesicles are found scattered in the cortical zone. In the early oocytes, the Golgi complex and centrosome are closely associated.

**Absorption of compounds**

*Hirsch* et al., have discovered that, when iron sugar is fed to an animal, iron becomes absorbed on Golgi complex *(Kedrowsky)*. *Van Teel* has shown that Golgi systems also absorb compounds of copper and gold. *Kedrowsky* has shown that Golgi complex of *Opalina* can absorb bismutose (compound of albumin and bismuth) and protargol (compound of albumin and silver). Thus, *Kirkman* and *Severing-haus* state that Golgi complex acts as a condensation membrane for the concentration: of products into droplets or granules.

**Plant cell wall formation**

The cell walls of plants are made up of fibrils which predominantly contain polysaccharides, along with some lipids and proteins. During cytokinesis a *cell plate is* formed between the two daughter nuclei, and has around it a membrane which later becomes the plasma membrane of the daughter cells. There is clear evidence that the polysaccharides are formed in the Golgi complex and transferred to the new cell wall which is laid down while the cells are still growing.

Substances like pectins and hemicelluloses, which form the matrix of the cell plate separating the plasma membranes are also contributed by the Golgi complex.

**Formation of intracellular crystals**

In the marine isopod, *Limnoria ligmorum,* which is a irrowing form there are present midglands whose cells consist 'crystals. These range up to 30A° in length and 15A° thick. It is been proved that these crystals are formed by Golgi complex and are known to contain protein and iron. They are thout enclosing membrane and usually spheroidal in shape. They are concerned with the secretory activity.

**Milk protein droplet formation**

In the lactating mammary gland of mice are produced )tein droplets wich are related with Golgi complex. These droplets usually open on to the cell surface by the fusion of their enclosing membrane with the plasma membrane.

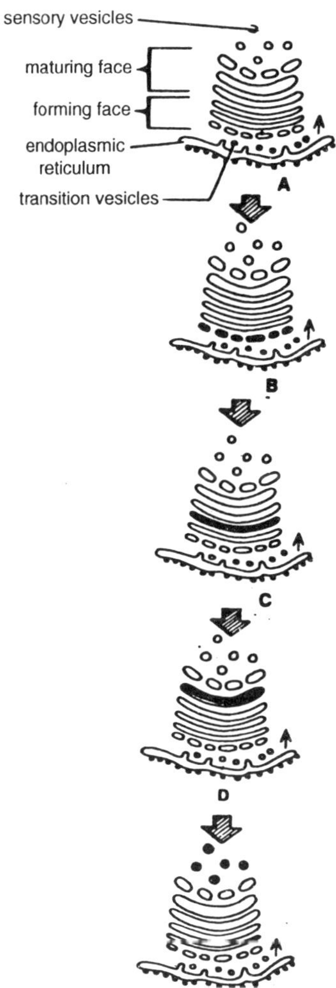


*Figure 5.5: Origin of dictyosome from endoplasmic reticulum from A to D and division to dictyosome—E, F, G.*

## Formation of lysosomes and vacuoles

Primary lysosomes are formed from the Golgi membranes the same way as the secretory vesicles. There is good evidence that dictyosomes accumulate hydrolytic enzymes in their more mature regions. Some vacuoles in plant cells have been found to contain small amounts of hydrolytic enzymes and these are presumed to have been derived from Golgi complex.

**Pigment formation**

In many mammalian tumour and cancer cells the Golgi complex has been described as the site of origin of pigment granules (melanin).

**Regulation of fluid balance**

A homology has been suggested between the Golgi complex and the contractile vacuole of lower Metazoa and Protozoa. The contractile vacuole expel surplus water from the cell. In certain Protozoa the Golgi complex is also concerned with regulation of fluid balance.

## ORIGIN OF GOLGI COMPLEX

Three different sources have been proposed from which new Golgi complex may arise:

**From endoplasmic reticulum**

*Essner* and *Novikoff* and *Beams* and *Kessel* have proposed that the Golgi cisternae arise from the ER. The rough endoplasmic reticulum after synthesizing specific proteins loses ribosomes and changes into smooth ER. Small transitory vesicles pinch off from smooth ER. These migrate to dictyosome. On reaching the forming face of dictyosome these fuse to form new cisternae and thus contribute to its growth.

By the fusion of these vesicles new cisternae are formed ,continuously on the forming face and on the maturing face the old cisternae break down into secretory vesicles. Thus Golgi exhibit a phenomenon of 'membranous flow.

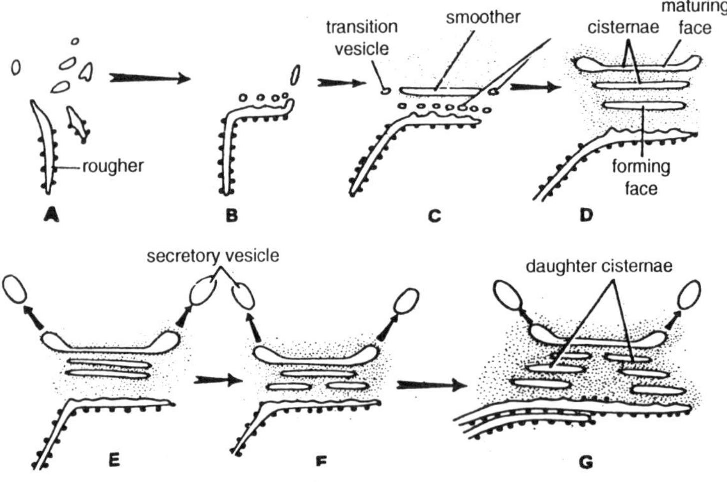


*Figure 5.6: Origin of Golgi complex by the division of pre-existing one.*

**From nuclear membrane**

*Bouch* described the origin of Golgi from outer membrane of nuclear envelope in brown algae. Vesicles are pinched off from outer nuclear membrane which fuse to form cisternae on the forming face of dictyosome.

Presence of zones of exclusion in relation with smooth ER or nuclear membrane, the occurrence of zones of exclusion in. dormant seeds of higher plants and the formation of ditcyosome from these zones in germinating seeds provide evidence in support of the above two theories about the origin of dictyosome.

**By the division of pre-existing dictyosome**

It has been observed that during cell division in both plants and animals, the number of dictyosomes increases and the number of dictyosomes in each daughter cell just after division is almost equal to the number in the parent cell prior to division. From this and other direct observations on the dividing cells it has been presumed that dictyosomes also divide during cell division.

# 6

# Endoplasmic Reticulum

Endoplasmic recticulum was first or all observed in 1945 by *Porter*, *Claude* and *Fullam*. They noted the presence of a network or recticulum of strands associated with vesicle-like bodies in the cytoplasm of the cultural fibroplast or thinlyspread tissue culture cells. Further electron microscopy by *Porter* and *Thompson* has revealed that these strands of reticulum are vesicular bodies inter-connected, so as to form a complex network in the inner endoplasmic part of the cytoplasm.

As this network is concentrated in the endoplasm of the cell more than in the ectoplasm, therefore, it is known as *endoplasmic reticulum* (*ER*), or *ergastoplasm* or *vaculolar system* of the cell. The endoplasmic reticulum is not visible in the cytoplasm of a living cell under the phase-contrast microscope but the observations by electron microscope on the fixed cell have revealed this structure. In 1952, studies with electron microscope further confirmed the presence of endoplasmic recticulum as reported by *Porter* and his colleagues. Recent studies have further confirmed and accepted the concept of a structural organisation of cytoplasm. Recently, under the phase-contrast microscope, *Fawcett* and *Ito*, and *Rose* and *Pomerat* have studied the structure and distribution of endoplasmic reticulum in the living tissue-culture cells.

### Occurrence

The endoplasmic reticulum occurs in all the eukaryotic cells except erythrocytes (R.B.Cs) of mammals. It is absent in prokaryotes. Its development varies considerably in various cell types. It is small and undifferentiated in eggs and in undifferentiated embryonic cells. Only a few vacuoles are present in the spermatocytes and muscle cells. However,

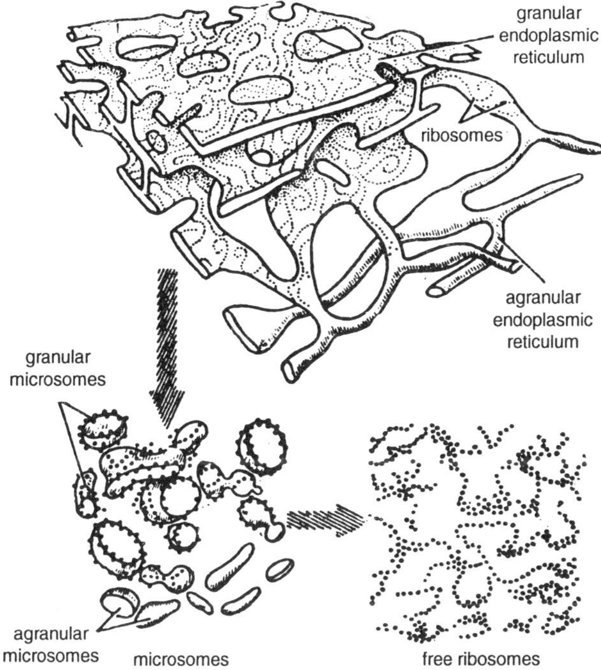


*Figure 6.1: Three dimensional structure of endoplasmic reticulum showing microsomes and ribosomes.*

it is highly organised in cells synthesizing proteins or in cells that are engaged in liopid metabolism.

## MORPHOLOGY OF ENDOPLASMIC RETICULUM

The endoplasmic reticulum has been found in all kinds of mature cells except the mature mammalian erythrocyte, which is also devoid of a nucleus. Actually the first description of these structures seemed with electron microscope by *Porter, Claude,* and *Fullam* in 1945 in cultured cells. These are membrane bounded sacs in the form of double membranes (cisterane) by *Sjostrand* or the name *cisternae* was given by *Sjostrand* and the name *tubules* was given by *Kurosumi*. Rounded and irregular *sacs* or *vesicles* were observed by *Weiss* 1953.

Morphologically the endoplasmic reticulum is composed of following three kinds of structures, viz., 1. cisternae, 2, vesicles and 3. tubules.

**Cisternae or lamellae**

They are long, flattened and usually unbranched tubules, which are arranged in parallel arrays. They are of uniform width throughout and their thickness varies from 40-50 m$\mu$. This pattern of reticulum is characteristic of basophilic regions of the cytoplasm and of those cells, which are active in protein synthesis. Lamellae or cisternae occur in the liver-cells, plasma-cells, pancreatic cells, brain-cells and in notochord cells etc.

**Tubules**

The tubules are small, smooth-walled, branched tubular spaces having a diameter of about 50-190 m$\mu$. These occur in cells that are busy in the synthesis of steroids like cholesterol, glycerides hormones. These are haphazardly arranged in the cytoplasm of developing spermatids of guinea pig, muscles cells and other non-secretory cells.

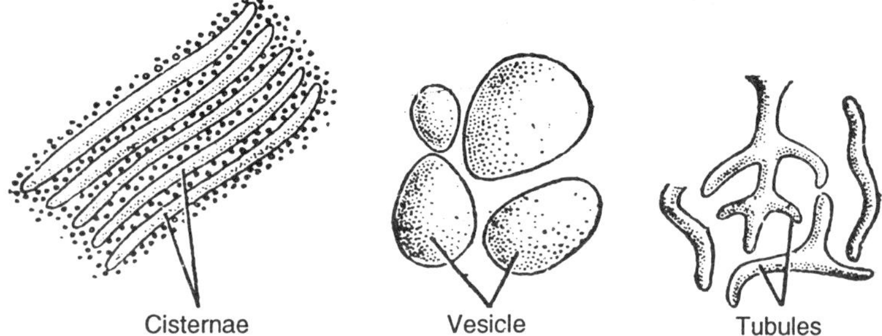


*Figure 6.2: Various components of the endoplasmic reticulum.*

**Vesicles**

The vesicles range in diameter from 25 to 500 $\mu$m and are for the most part rounded in shape. These are abundant in the cells engaged in the protein synthesis as in hepatic and pancreatic cells.

All these three patterns of endoplasmic reticulum may occur in the same cell or in different cells. Their arrangement also liffers in different cells viz., in parallel rows in the li\ er-cells of mammals; haphazardly in pancreatic cells or in the form of a iet-work of tubules in striated muscle-cells. In notocbordal cells of *Ambyostoma* larva, the pattern of cisternae is of still another type.

**Ultrastructure of endoplasmic reticulum**

All the three structures of the endoplasmic reticulum are bounded by a thin membrane of 50 to 60 A° thick. Like the plasma membrane, nucleus, etc., its membrane is also formed of three layers—the outer and inner dense layers are composed of protein molecules, and the two middle thin and transparent layers are of phospholipids. The endoplasmic reticular membrane is continuous with the plasma membrane, nuclear membrane and membrane of Golgi complex. The lumen of the endoplasmic reticulum acts as a passage for the secretory products and *Palade* has observed the secretory granules in it.

**Types of endoplasmic reticulum**

On the basis of presence or absence of ribosomes they are of two types:

*(i) Granular or Rough Walled Endoplasmic Reticulum.* When the particles or *ribosomes* are present on the wall of E.R., it is called rough walled E.R. These particles are always present at the outer surface of the ER. *i.e.* on the surface of the limiting membrane facing the continuous phase the matrix of the cytoplasm. The elements (E.R.) with rough surfaces are high in ribonucleic acid and are intensely basophilic. The membranes themselves are not rough, but associated with their outer surfaces are tiny particulate components 100 or 150A° in diameter.

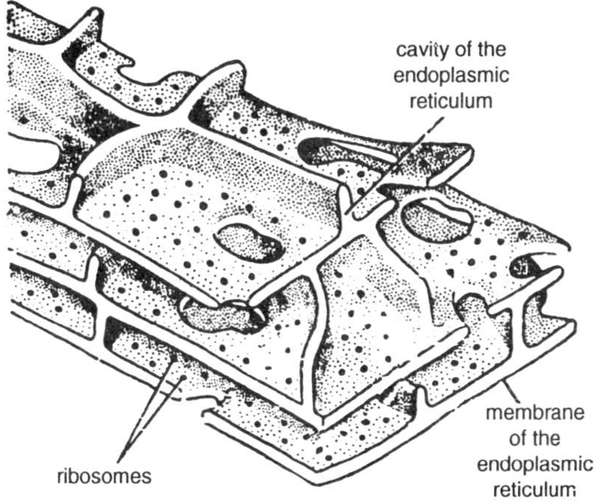


*Figure 6.3. Three dimensional structure of granular endoplasmic reticulum.*

These are called *ribonucleoprotein* (*RNP*) *particles* or *ribosomes* and contain as the average 40% RNA and 60% protein. The elements having ribosomes are usually of the cisternal type and are found in cells active in protein synthesis. Biochemical studies have indicated that the ribosomes are important in protein synthesis, even though the membranes are not always necessary for this activity. Other functions will be explained later.

The smooth surfaced E.R. is often continuous with the rough surfaced E.R. thereby making the absence or presence of ribosomes the only significant difference between the two. The continuity between smooth

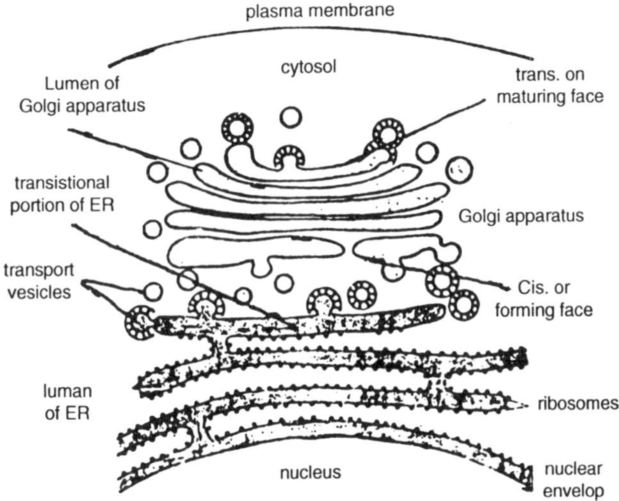


*Figure 6.4: Endoplasmic reticulum and other cell organelles.*

and rough E.R. has been repeated by demonstration. It has been suggested more over that one grows from the other but which from which is uncertain. The ribosomes can readily be dissociated from the endoplasmic reticulum membranes by treatment with deoxycholate.

**Table 6.1: Differences between rough and smooth ER.**

| | Smooth ER | Rough ER |
|---|---|---|
| 1. | Well developed in steroid hormone secreting cells. | Well developed in protein secreting cells. |
| 2. | It tends to be tubular. | It tends to be cisternal. |
| 3. | It is less stable and autolysis readily. | It is com-aratively more stable. It may persist for sometime after the death of the cell. |
| 4. | It is devopd of ribosomes | Ribosomes are found associated. |

*(ii) Smooth Walled Endoplasmic Reticulum.* The name smooth walled is given to that portion of endoplasmic reticulum that is devoid of ribosomes. Like the rough walled endoplasmic reticulum smooth form shows a characteristic morphology which is tubular rather than cisternae. These smooth walled endoplasmic reticulum is found in the cells that are active in the synthesis of steroid compounds such as cholesterol, glycerides, and the hormones (testosterone and progesterone). It is studied by *Fawcett* that are also present in the pigmented epithelial cells of the retina which are involved in the metabolism of vitamin A in the production

of visual pigment. Glycogen storing cells of the liver contain the smooth, tubular elements of the endoplasmic reticulum.

## MODIFICATION OF ENDOPLASMIC RETICULUM

### Sarcoplasmic reticulum

Sarcoplasnr,ic reticulum, found in the skeletal and cardiac muscles is a highly modified form of smooth ER. It was first reported by *Veratti* as delicate plexuses in skeletal muscles surrounding the myofibrils. Electron microscopy showed it to be composed of a network of membrane like tubules which run longitudinally in the interfibrillar sarcoplasmic space for the length of each sarcomere.

At the level of H and I bands, these tubules merge with large cisternal structures. At the H band level this cisterna, called the central cisterna, forms a sieve-like structure round the myofibrils. At the level of I band, these tubules merge with the large terminal cisternae, from which transverse tubules extend peripherally to the sarcolemma and are continuous with and are deep invaginations of it. It is generally believed that sarcoplasmic reticulum plays a role not only in distributing energy-rich material needed for muscular contraction but also in providing the necessary channels for transmitting impulses along the surface and conveying the action potential from the surface to the myofibrils within.

### Ergastoplasm

There are certain regions in the cytoplasm that stain with basic dyes. To these regions various names have been given like chromidial substance, basoplasm, ergastoplasm and so forth. The term *ergastoplasrn* was given by *Garnier* in 1899 to those cytoplasmic filaments in the cells of exocrine glands which stained readily with basic stains. *Weiss* (1953) referred to the cisternal elements as ergastoplasmic sacs. In nerve cells, such areas are called *Nissil bodies*. Electron microscopic studies reveal it to be an accumulation of ribosomes situated on the parallel lamellae of ER stacks of accumulated freely in the groundplasm. Studies of *Caspersson*, *Brachet* and others have demonstrated that the basophilic nature of ergastoplasm is due to the ribonucleic acid. Smooth E.R. areas of cytoplasm are never ergastoplasm.

### Isolation of Endoplasmic Reticulum

Endoplasmic reticulum, can also be isolated mechanically with the help of centrifuge. When tissues or cells are disrupted by homogenisation the E.R. is fragmented into many smaller closed vesicles called microsomes (100 nm diameter), which are relatively easy to purify.

Microsomes derived from rought E.R. are studeid with ribosomes and are called rough microsomes. Many vesicles of size similar to that of rough microsomes, but lacking attached ribosomes, are also found or these homogenates. Such smooth microsomes are derived in part from smooth portions of the ER and in part from vesiculated fragments of plasma membrane, Golgi complex and mitochondria (the ratio depending on the tissue). Thus, while rough microsomes can be equated with rough portions of ER, the origins of smooth microsomes cannot be easily assigned. An outstanding exception is the liver. Because of the exceedingly large quantities of smooth ER in the hepatocyte, most of the smooth microsomes in liver homogenates are derived from smooth ER.

Ribosomes, which contain large amounts of RNA, make ah microsomes more dense than smooth microsomes. As a result, the rough and smooth microsomes can be separated fron each other by sedimenting the mixture to equilibrium in sucrost density gradients. When the separated rough and smoott microsomes of a tissue such as liver are compared with respect to such properties as enzyme activity or poly peptide composi. tion, they are remarkably similar, although not identical. It, therefore, seems that most of the components of the ER membrane can diffuse freely between rough and smooth regions of the E.R. membrane, as would be expected for a fluid, continuous membrane system.

## ORIGIN OF ENDOPLASMIC RETICULUM

### Multi Step Mechanism

Although the origin of the new endoplasmic reticulum membrane has not been fully understood. The views are there. In fact, one of the possible function attributed to the endoplasmic reticulum is that of membrane biosynthesis. The protein components of endoplasmic reticulum and other membranes may be assembly by activity of the endoplasmic reticulum. There is certainly convicing evidence that Golgi membranes and many cytoplasmic vesicles can be derived from endoplasmic reticulum.

Moreover, endoplasmic reticular membranes appear to be continuously synthesized, having a relatively high rate of turn over. At the same time, the several elements of endoplasmic reticulum in the cell are asynchronous in this respect, they are not all replaced at the same time or with the same rate. It has also been suggested that membranes of the endoplasmic reticulum are formed not from pre-existing elements but from he ground substance of the cytoplasm. Thus the process by which a membrane is modified chemically and structurally is called membrane differentiation.

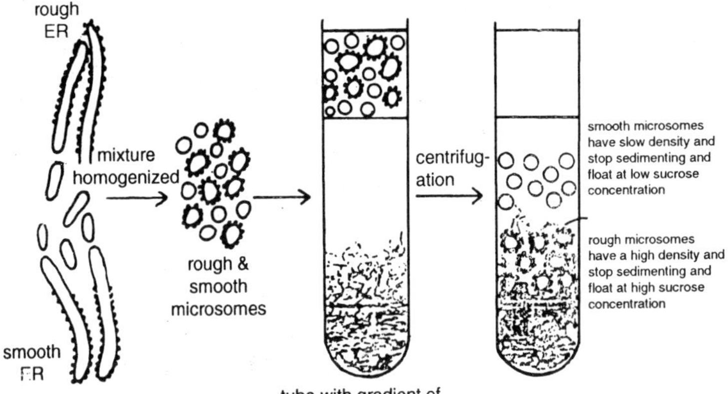


*Figure 6.5: Isolation procedure used to purity RWER & SWER.*

**From Nuclear membrane**

The vaculoes derived from the evagination of the outer membrane of the nuclear envelope, which separates from its inner partner, leaving cavities between. Shortly after the separation, small vesicles appear near the nuclear envelope, suggesting that parts of the envelope give rise to elements of the endoplasmic reticulum. Thus the endoplasmic reticulum seems to have its origin in the nuclear envelop in undifferentiated cells.

## ENZYMES OF THE ER MEMBRANES

The membranes of the endoplasmic reticulum are found to ontain many kinds of enzymes which are needed for various important synthetic activities. The most important enzymes are the stearases, NADH-cytochrome C reductase, NADH diaphorase glucose-6-phosphotase-and-$Mg^{++}$ activiated ATPase. Certain enzymes of the endoplasmic reticulum such as nucleotide diphosphate are involved in the biosynthesis of phospholipid, ascorbic acid, glucuronide, steroids and bexose metabolism. The enzymes of the endoplasmic reticulum perform the following important functions

1. Synthesis of glycerides, e.g., triglycerides, phospholipids, glycolipids.
2. Metabolism of plasmalgens.
3. Synthesis of fatty acids.
4. Biosynthesis of the steroids, e.g., cholesterol biosynthesis, steroid hydrogenation of unsaturated bonds.
5. $NADPH_2$ + $O_2$—requiring steroid transformations: Aromatis-

atiop and hydroxylation.

6. $NADPH_2$ + $O_2$—requiring steroid transformations : Aromatic, hydroxylations, side-chain oxidation, deamination, thioether oxidation, desulfuration.
7. L-ascorbic acid synthesis.
8. UDP-uronic acid metabolism.
9. UDP-glucose dephosphorylation
10. Aryl-and steroid sulphatase.

**Functions of the Smooth ER**

It is a little artificial to try to separate the activities of the ;rough and smooth ER because of their numerous interconnections. One obvious difference, of course, is participation of the rough ER in protein synthesis by virtue of its associated ribosomes. Smooth ER, on the other hand, is more involved with lipid synthesis.

Cells concerned largely with synthesis of lipids have welldeveloped smooth ER just as cells that synthesize and secrete proteins have well developed rough ER. Thus, cells of the adrenal cortex (which synthesize steroid hormones) have abundant smooth ER. So do intestinal absorptive cells, for the smooth ER supports triglyceride synthesis from the mixture of fatty acids, monoglycerides, and diglycerides taken up by these cells from the intestine. (Dietary fat is broken down by increatic lipase to provide these raw materials.) Fat drops (cbylomicrons), comprised almost exclusively of triglyride, can be readily identified within the smooth ER near thetestinal lumen shortly after a meal.

The smooth ER of liver cells has functions that, while not ique to the liver, are certainly more prominent there. These fictions include regulation of glycogen breakdown (via icose 6-phosphatase, an ER-associated enzyme), and the njugation and oxidation of a variety of naturally occurring d foreign substances, the purpose of which is usually to 'minate their biological activity. When applied to drugs, this )cess is referred to as *detoxification.*

The reactions that detoxify substances consist of oxidaons reductions, hydrolyses, or a covalent linking (conjugaon) to soluble small molecules, particularly conjugation to lucuronic acid, a sugar derivative. These changes either activate the substance or make it more soluble and hence lore readily eliminated by the kidreys. A wide variety of atural and artificial chemicals are affected in this way, incluing environmental pollutants and many drugs. Birth contral pills, for example, were made possible by the discovery of drugs having some of the activities of estrogens and progesterone but which are not rapidly inactivated by the liver as are the natural steroid hormones.

In general, these reactions of the smooth ER serve a valuable protective function and also play a role in the normal handling of fatty acids, bile salts, steroids, and heme recovered from hemoglobin breakdown. The responsible enzymes are associated with endoplasmic reticulum in several different cell types, and related systems have been identified even in prokaryotes. They are especially prominent, however, in liver.

Most of the biotransformations in this category are oxida tions, primarily because of the multitude of ways in which organic molecules can be oxidised. The first key to understanding the mechanism was a proposal by Howard S. Mason of the University of Oregon Medical School for the existence of "mixed function oxidases" utilising both NADPH and molecular oxygen.

The principal enzyme is cytochrome P450, a heme protein so named because it absorbs light maximally at 450 nm when in the reduced form. This protein is capable of binding substrate and $O_2$, passing one of the oxygen atoms of $O_2$ to the substrate in the form of an OH group (an oxidation of the substrate) while the other oxygen atom combines with $H^+$ to form water:

$$S + P450(Fe^{3+}) \xrightarrow{O_2 + 2e^-} S - P450(Fe^{2+}) - O_2^- \xrightarrow{2H^+} P450(Fe^{3+}) + S - OH + H_2O$$

Here, S is the substrate and $Fe^{3+}$ the heme-associated iroliof P450. The electrons are delivered to P450 mostly from NADPH by an endoplasmic-reticulum-associated electron transport chain involving flavoproteins and in some cases another cytochrome known as cytochrome $b_5$.

Cytochrome P450 and other elements of the *microsomal electron transport chain* seem to be membrane-bound proteins present mostly on the cytoplasmic surface of the smooth endoplasmic reticulum. Their concentration, however, varies with need.

As mentioned earlier, phenobarbital can cause the amount of smooth ER to more than double, but the amount of P450 per cell may at the same time increase fivefold. Although there are apparently several versions of P450 with somewhat different specificities, there is broad overlap so that induction with one substrate increases the ability of the ER to detoxify other substrates handled by the same system even though the other substrates may not themselves be effective inducers of P450.

Thus, for example, treatment of a human with more than on ER-metabolised drug at a time often requires careful adjustment of dosage.

**Functions of the Rough ER**

It has long been assumed that proteins destined for secrestion from the cell are synthesized on ER-bound ribosomes while cytoplasmic proteins are translated for the most part on free ribosomes. This separation is not hard and fast, for even cells that secrete little or no protein typically have significant amounts of rough ER. However, when radioactive amino acids are injected into a cell that makes secreted proteins, the radioactivity very quickly becomes recoverable in the rough microsomal fraction, thus supporting out generalisation.

What is more, further investigation has revealed that this rough ERassociated protein is immune to proteases. Not only does translation occur on membrane-bound ribosomes, but the secretory protein, instead of passing into the cytoplasm, appears to pass instead into the cisterna of the rough ER, and hence into microsomes where it is protected from experimentally added proteases.

From the rough ER, secreted proteins pass sequentially to mooth surfaced ER, Golgi apparatus, secretion vesicles, and nally to the exterior of the cell. The questions to be ansrered here are (1) how does mRNA of the secretory proteins ind specifically to membrane-bound ribosomes, and (2) how oes the protein get from the ribosome to the interior of the R ? The answers to those critical questions appear to be at and, thanks in large part to pioneering experiments performed y Gunter Blobel at Rockefeller University and David Saba.tini t New York University School of Medicine.

**Synthesis of Secretory Proteins**

The obvious first question is whether there is anything different about membrane-bound ribosomes. If they are the same as free ribosomes, then the specific association between mRNA and rough ER must be due to a feature peculiar to the mRNA or to the protein made from it. In fact, that is where we must look, because it appears that ribosomes are selected nonspeci fically for membrane association.

Electron microscopy reveals that membrane-bound ribo somes are attached by their large, 60S subunit, with the smal or 40S subunit sitting on top like a snowman. There is nov considerable evidence for the existence of specific binding site within the membrane capable of attaching only to the 60S subunit.

Evidence for specific ribosomal binding sites was obtained by experiments in which rough microsomes were stripped of their ribosomes and then exposed to free ribosomes obtained either in this way or from the cytoplasmic pool.

The result was reattachment of ribosomes to roughly the original concentration. Equivalent treatment of smooth microsomes results in much lower rates of ribosomal attachment, though the affinity of the attached ribosomes is comparable, leading one to conclude that the same type of binding site is involved but present in much lower concentration. Control experiments with plasma membranes and Golgi membranes showed weak interaction with ribosomes and low capacity for attachment. Still further evidence for specific binding sites was obtained in experiments where the stripped microsomes were exposed to very mild proteolytic activity, destroying their capacity to accept ribosomes by damaging the ribosomal binding site.

The presumed ribosomal binding sites (or receptors) appear as membrane proteins on freeze-fracture, extending well into and possibly through the lipid bilayer. These proteins, like most other membrane proteins, enjoy considerable feeedom to diffuse laterally, so that "patching" and even a sort of "capping" of bound ribosomes can occur *in vitro* at temperatures above the phase transition of the memberane. This mobility of course, would facilitate formation of the polysome and probably translation, which requires that the mRNA and ribosome move with respect to each other. Absolute freedom of movement is apparently not allowed *in vivo*, however, otherwise the distinction between rough and smooth ER would disappear.

It is probable, based on other experiments, that the association between mRNA and ribosomes-i.e., polysome formation -begins while ribosomes are free. They later attach to the ER, depending on availability of ribosome binding sites. This association, according to some data, is fostered by binding of' the mRNA to membrane ; other data point to association between the new polypeptide chain and membrane. Both, factors may be important.

Work from Sabatini's laboratory demonstrated that the 3' end of mRNA from membrane-bound polysomes can remain attached to the membrane even after the ribosomes have been dissociated and stripped away. Other experiments suggest that certain mRNA can associate with microsome formed from stripped rough ER even in the absence of ribosomes. The 3' end of mRNA is of course, the end that gets translated first.. It is also the end carrying, in eukaryotic cells, a 1 50-200 nucleotide segment of poly A. However, the poly A is found on mRNA of both types, free as well as membrane bound. Hence, it is probable that mRNA ending for secreted proteins contains near this segment a segquenc that binds specifically to ER-or, more likely, to protein receptors within the ER.

Another line of investigation, carried out in large part by obel and his colleagues, points to an affinity between memane and an N-terminal segment of those polypeptide chains stined for secretion. Translation of the immunoglobulin ;ht chain, for instance, produces first a 20- amino-acid segcnt that is later cleaved from the chain. This segment is largely hydrophobic in character, giving it an affinity for Imbrane.

Comparable segments, also later sacrificed, have en identified at the N-terminus of other incomplete secretory oteins. Removal of these segments, at least in the case of the immunoglobulin light chain, is catalysed by enzymes of the rough ER and takes place even before completion of the polypeptide chain. These two proposals for directing secretory protein mRNA to rough ER-namely, attachment of the mRNA itself, and affinity between the nascent (incomplete) polypeptide and membrane—are not mutually exclusive. Both may function to help ensure translation of these mRNAs specifically at the rough ER.

**Isolation of Secretory Proteins by the ER**

The next problem is how the polypeptide chain that is destined for export gets into the cisternae of the ER. It does so very quickly, as demonstrated by Sabatini and Blobel in 1970, when they showed that mild proteolysis of rough microsomes did not destroy the nascent polypeptide chains but divided them into two parts. The incomplete C-terminal segment was protected by the ribosome on which it was being synthesized. The N-terminal segment was protected by the membrane of the ER. Hence, passage into the ER cisterna takes place during translation, leaving only a small segment exposed to the cytoplasm at any one time.

How the polypeptide chain gets through the lipid bilayer is not so clear, but it is quite reasonable to propose that the membrane protein serving as a ribosomal receptor also has a channel through its core that opens into the cisterna of the ER. The channel would have to be no larger than some of the water and cation pores associated with certain proteins of the plasma membrane discussed earlier, for there is free rotation about most of the nonpeptide bonds in the peptide chain. Hence the chain has great flexibi\ity, permitting the amino acids to snake their way single file through the proposed pore.

Folding into secondary and tertiary structures also begins while the polypeptide is still in the process of synthesis. Formation of such structures would prevent return of the peptide through the narrow orifice from hence it came, thus trapping it. in the cisternae of the rough ER.

**Glycosylation of Proteins**

Nearly all proteins destined for secretion are glycoproteins. notable exception is albumin, actually the most common )tein in serum.) Glycosylation—i.e., the addition of carbodrate—begins even before the polypeptide chain is complete. The sugar molecules appear to be added one at a time, insferred usually from the nucleotide UDP, which serves as carrier. Glycosylation is ordinarily still in progress when e polypeptide chain is completed and released into the cisterna of the rough ER. However, if the glycoprotein is to ntain a terminal galactose, fucose, or sialic acid, those sugars e added in the Golgi apparatus where the appropriate sugar tnsferases are localised.

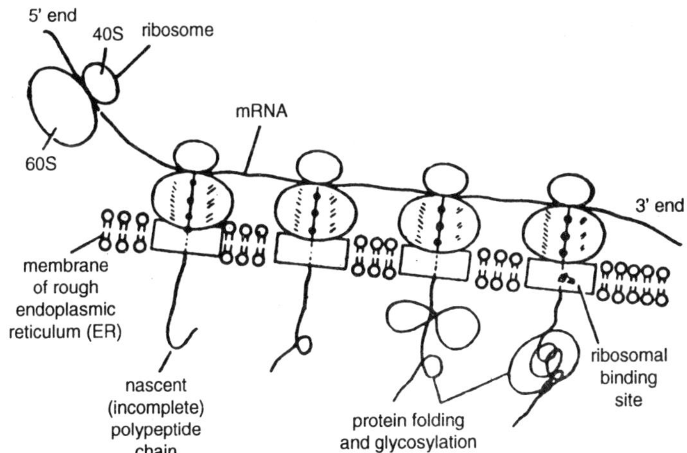


*Figure 6.6: Synthesis of secretory proteins on rough endoplasmic reticulum*

The stepwise addition of sugars to proteins from a UDPgar intermediate is well documented. However, in recent ars it has become clear that animal cells like cells of bacteria A plants, have a second and quite different way of glycosy:ing proteins. In this second pathway, the sugar chains are sembled in a stepwise manner from nucleotide-sugars, but the sembly is not on the protein itself. Rather, it is on a lipid, either *retinol (vitamin A)* or on one of a family of lipids called *dolichols*. The latter system has been particularly well studied.

In the lipid-linked assembly of certain oligosaccbarides, transfer is initially to dolichol phosphate, creating a highenergy pyrophosphate (—P—O—P—) linkage. The first sugar in the chain is typically N-acetylglucosamine (GlcNAc). Thus we have (where P is phosphate)

UDP—GlcNAc+dol—P → dol—P—P—GlcNAc+UMP

Thereafter the chain is lengthened one sugar at a time. When the chain is nearly complete, it is transferred intact to the protein, using cleavage of the pyrophosphate to drive the reaction.

Why should two so very different mechanisms for protein glycosylation exist? All the evidence is not in but it appears probable that the lipid-linked system functions in the *glycosylation of membrane proteins.* The assembly of carbohydrate chains onto dolichol seems to take place, at least in part, on the cytoplasmic side of the membrane, rather than in the cisternae of the ER.

Although membrane proteins seem to be synthesized largely on membrane-bound ribosomes (one season for the presence of rough ER in nonsecretory cells), one can imagine that their hydrophobic nature causestibemto become incorporated into the membranes of the endoplasmic reticulum rather than confined to the cisternae within. Transfer from dolichol might then take place at the surface of the ER, which would also explain how membrane proteins come to be glycosylated in such an asymmetric way—i.e., on only one side. One of the functions of carbohydrate on secretory proteins, then, might be to provide a hydrophilic coating to help keep the proteins in the cisternae of the ER.

In any case, membrane proteins, like those destined for export, pass from the rough ER to smooth surfaced ER adjacent to the Golgi apparatus and then ordinarily to the Golgi apparatus itself where terminal sugars may be added in a stepwise fashion. That organelle, the Golgi apparatus, is our next topic.

## OTHER FUNCTIONS OF E.R.

Many functional interpretations of endoplasmic reticulum based on the polymorphic aspects of its components in a iety of cells and its different stages of activity. More reliable interpretations are based on the isolation studies just itioned. The following functions are based on the same I known facts together with hypothesis.

### Mechanical Support

ER contributes to the mechanical support of the cytoplasm )y dividing the fluid contents of the cell into compartments. this makes possible the existence of ionic gradients and elecrical potentials along ER membranes. This concept has been pecially applied to sarcoplasmic reticulum.

### Exchange of ions and any other fluid

The membranes of the endoplasmic reticulum may regulate the

exchange between the inner compartment or cavity and the cytoplasmic matrix. The following statistic given an impressive idea of the surface area available for exchange : 1 gm of liver contains about 8 to 12 square meter of endoplasmic reticulum. After isolation, microsomes expand or shrink according to the osmotic pressure of the fluid. Diffusion and active transports may take place across the membrane of the endoplasmic reticulum.

**Intracellular Circulation**

The endoplasmic reticulum may act as a kind of circulatory ystem for intracellular circulation of various substances. vembrane flow may be an important mechanism for carrying )articles, molecules and ions into and out of the cells by way & vascular system. The "pinocytosis", a "cellular drinking" ilso takes place by endoplasmic reticulum.

By this mechanism, particles attached to the surface of the ;ell or suspended in the fluid medium can be incorporated into he cytoplasm. The similar mechanism but working in a reverse lirection can effect the transport of a particle from the interior of the cytoplasm to the outer medium. The continuities observed in some cases between the endoplasmic reticulum and the nuclear envelope suggest that the membrane flow may also be active at this point. This flow would provide one of the several mechanisms for export of RNA and nucleoproteins from the nucleus to the cytoplasm.

**Protein Synthesis**

Proteins may be synthesized to be utilised within the cell or these may have to be exported outside the cell to the site of their utility. It is the latter kind of proteins in whose synthesis, endoplasmic reticulum plays an important role. For instance, rough endoplasmic reticulum, which has attached ribosomes carries synthesis of secretory proteins on these ribosomes and export them. Synthesis of tropocollagen, serum proteins and secretory granules are some examples of secretory proteins.

The protein molecules synthesized on attached ribosomes are discharged and penetrate into the cavity of ER, where they are stored or exported outside. During the transport of these products, three types of membranes ER-Golgi membraneplasma membrane should interact and remian interconnected or disconnected due to fusion and fission respectively.

**Synthesis of Lipid**

The cells in which active lipid metabolism takes place are found to contain large amount of the smooth type of endoplasmic reticulum.

According to some workers as *Christensen* and *Claude* the smooth type of endoplasmic reticulum is related with the synthesis and metabolism of the lipids.

**Synthesis of Glycoge**

The smooth endoplasmic reticulum of the glycogen storing cells of the liver and the cells of certain plants is found to be associated with the synthesis, storage and metabolism of the glycogen. But, *Porter* and *Peter* have suggested that smooth type of endoplasmic reticulum is related to glycogenolysis (digestion of glycogen) and not to glycogenesis (synthesis of the glycogen).

**Detoxification**

Smooth ER is also involved in the detoxification of many endogenous and exogenous compounds. Prolonged administra:ion of certain drugs (phenobarbitol) results in the increased tetivity of enzymes related to detoxification, as well as other enzymes, and a considerable hypertrophy of the SER. This is tlso applicable to administered steroid hormones.

**Synthesis of Cholesterol and Steroid Hormones**

Cholesterol is an important precursor of steroid hormones. Fhe major site of cholesterol synthesis is the ER. In liver cells he SER is believed to be concerned with both the synthesis tnd storage of cholesterol.

In the testis, ovary and the adrenal cortex the SER has a role in the synthesis of steroid hormones. The enzymes catalysing biosynthesis of androgens have been located in the SER. There is a strong correlation between the amounts of SER in cells and the capacity to synthesize steroid hormones.

**Amphibian Development**

There are evidences to suggests that the ER contributes in several ways to the development of the amphibian embryo.

**Cell Differentiation**

Some specific instances of development have been studied in detail which more or less confirm the contention that the ER is important in the process of cell differentiation. Not only this much, ER also plays role in coordinating the differentiation.

**Formation of Microbodies**

Closely related with the ER are microbodies, which are small granular bodies filled with an electron dense substance and limited by a single membrane. Microbodies are formed as dilations of the ER and frequently show connections with the ER cisternae. They are rich in the enzymes peroxidase (and are hence. also called *peroxisomes),* catalase, and D-amino acid oxidase. In plant cells the enzymatic content is different and the boides are called *glyoxy-somes* because they, include enzymes of the glyoxylate cycle.

**Enzyme Activities and Cellular Metabolism**

Numerous enzymes mainly those involved in the metabolism of steroids (cholesterol and glycerides), phospholipids and hormones (testosterone and progresterone) are associated with the membranes of smooth endoplasmic reticulum. These membranes provide an increased inner surface for various metabolic reactions and they themselves take an active part in them by means of attached enzymes. This facilitates free union of enzymes with their substrates.

**Role of Endoplasmic Reticulum in Intra-Cellular Impulse Conduction**

The existence of endoplasmic reticulum separating the cytoplasm into two compartments makes possible the existence of ionic gradients and electrical potentials across these intracellular membranes. The idea has been applied to the sacroplasmic reticulum, a specialised form of smooth surfaced endoplasmic reticulum found in striated muscles which is now being considered as intracellular conducting system. On the basis of some evidences, it has been postulated that the sarcoplasmic reticulum transmits impulses from the surface' membrane into deep regions of the muscle fibres.

**Formation of Plasmodermata**

Electron microscopic studies suggest that the endoplasmi reticulum in plants plays a special role in the interconnectio of cells through the cytoplasmic strands called *plasmodermata.*

**Role of E.R. During Cell Divisior**

During cell division, some of the elements of reticulun contribute in the formation of the new nuclear membrane after karyogamy. The nuclear membrane breaks up into fragment; in the early part of the division which finally disintigrate intc small vesicles. These vesicles move towards the pole of the spindle as the metaphase starts, where they are indistinguishable from the elements of ER. From the polar ends of the cell, elements of ER as well as the fragmented vesicles migrate into the regions around the chromosomes, which are grouping at the poles. Most of these elements of ER join or fuse around each group of daughter chromosomes to form a new nuclear envelop.

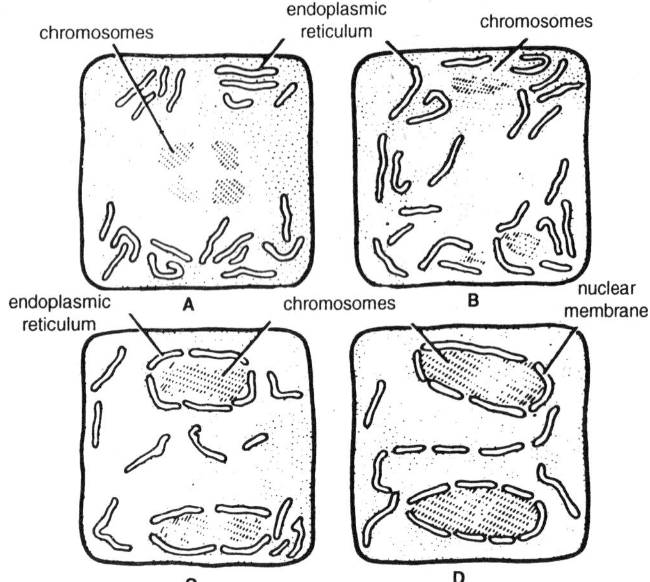


*Figure 6.7: Diagrammatic representation of formation of nuclear membrane form endoplasmic reticulum.*

## Transportation of Message from Genetic Material

ER provides passage for the genetic material to pass from the nucleus to the various organelles in the cytoplasm, thereby controlling the synthesis of proteins, fats and carbohydrates.

### ATP Synthesis

ER membranes are the sites of ATP synthesis in the cell. The ATP is used as a source of energy for all the intracellular netabolism and transport of materials.

### Formation of Cell Organelle

Most of cell organelles like Golgi complex, mitochondria, lysomes, nuclear membrane and cell plate etc. are usually developed from endoplasmic reticulum.

# 7

# LYSOSOMES

The concept of the lysosome originated from the development of cell fractination techniques by which different subcellular components are isolated. By 1949 a class of particles having centrifugal properties some what intermediate between those of mitochondria and microsomes was isolated by *de Duve* and found to have a high content of acid phosph,atase and other hydrolytic enzymes. Because of their enzymatic properties they were named lysosomes (Gr., *Lysis*= disolution, *soma*= body). According to *Gohan* (1972) membrane bounded storage granules containing digestive enzymes are considered lysosomes of plant Alls. Hence *spherosomes, aleurone granules* and *vocuoles* of plant cells are supposed to have lysosome like functions.

## OCCURRENCE

With the exception of mammalian R.B.C., the lysosomes have been reported practically from all the animal cells. The presence of lysosomal particles have also been suspected (and in some cases established) in protista (protozoan, slime-moulds, fungi, algae and prokaryotic protista). In plant cells, considering the evidences as a whole, there now seems little doubt about their presence. Further they have strong affinities with the lysosomes of animals and protista.

In plants, further they should not be confused with spherosomes in function. According to *Pitt* lysosomes and spherosomes are two different organelles and the later are comparable to lipid droplets of animals. *Yatsu* and *Jack* have clearly shown that spherosomes are morphologically distinct organelles. *Gahm* reviewed the occurrence and histochemistry of plant lysosomes.

# MORPHOLOGY

## Shape and Size

The shape and size of lysosome is variable. Morphologically they can be compared with *Amoeba* and white blood cells (W.B.C.). Due to their changing habit they cannot be accurately identified as the basis of their shape. Normally lysosomes vary in size from 0.4 to 0.8,u but they may be as large as 5$\mu$ in mammalian kidney cells and are exceedingly large in phagocytosis.

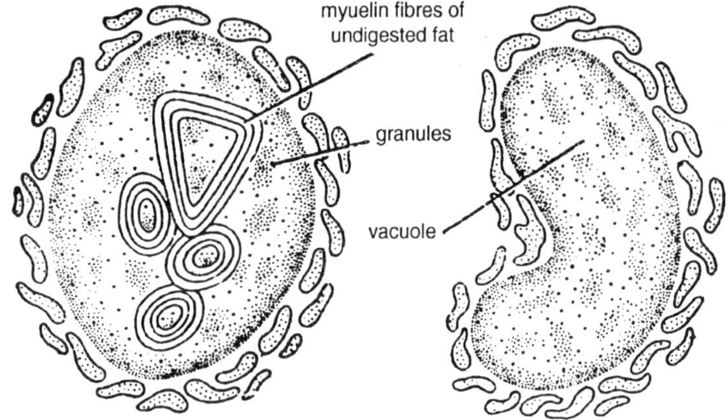


*Figure 7.1: Two types of lysosomes from the kidney cells of rat.*

## Structure of Lysosomes

Like other cytoplamic complexes, lysosomes are like round tiny bags filled with dense material and digestive enzymes. They consist of two parts (1) limiting membrane and (2) inner dense mass.

### *Limiting membrane*

This membrane is single, unlike that of mitochondria and composed of lipoprotein. The chemical structure is homologous with the unit membrane of plasma lemma consisting of bimolecular layer as suggested by *Robertson.*

### *Inner dense mass*

This enclosed mass may be solid or of very dense contents. Some lysosomes have a very dense outer zone and less dense inner zone. Some others have cavities of vacuoles within the granular material. Usually they are supposed to possess denser contents than mitochondria.

They show their polymorphic nature and their contents vary with the stage of digestion, as they help in intracellular digestion.

## Permeability of Lysosomal Membrane

The lysosomal membrane is impermeable to substrates of the enzymes contained in the lysosome. Certain substances, called *labilizers,* cause instability of the lysosomal membrane, leading to release of enzymes from the lysosome. Other substances, called *stabilisers,* have a stabilising action on the membrane. A list of some labilisers and stabilisers is given in Table.

**Table 7.1: Lysosomal labilisers and stabilisers.**

| Labilisers | Stabilisers |
|---|---|
| Vitamin A | Cholesterol |
| Vitamin B | Cortisone |
| Vitamin K | Cortisol |
| Vitamin E (high dose) | Vitamin E (low conc.) |
| Progesterone | Chloroquine |
| Testosterone | Phenothiazines |
| 13-estradiol | Antihistamines |
| Ubiquinone | Heparin Digitonin |
| Polyene antibiotics | |

A labiliser might increase the permeability of the lysosomal mbrane to small solutes like sucrose. The osmotic swelling ich results might completely disrupt the membrane.

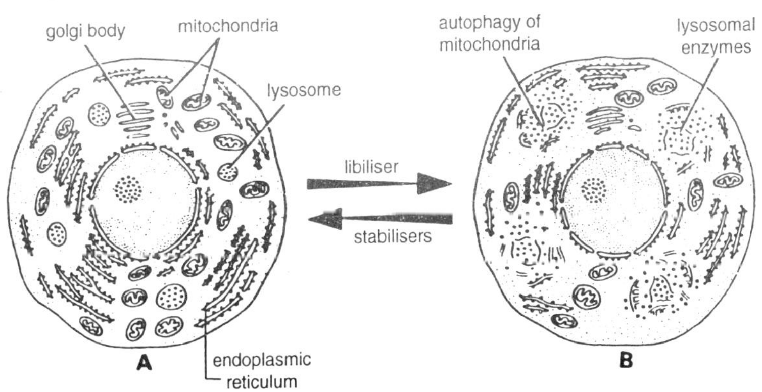


*Figure 7.2: Activities of the suicide bags in the cell.*

The limited permeability of lysosomal membrane explains why lysosomal hydrolases do not have a direct access to cellular components. This prevents uncontrolled digestion of the cell contents by the lysosomal enzymes.

**Polymorphism**

Lysosomes are polymorphic in nature. The polymorphic nature is due to variation in contents of lysosomes with different stages of digestion. Generally lysosomes can be traced in four forms given below

(1) *Primary lysosomes.* These are also called the *true, pure* or *original lysosomes,* having a single unit membrane containing enzymes in the inactive forms.

(2) *Secondary lysosomes.* These are also called the *phagosomes* as they contain the engulfed material and enzymes. The fused mass is called the secondary lysosomes. The enzyme present in such lysosomes gradually digests the engulfed material.

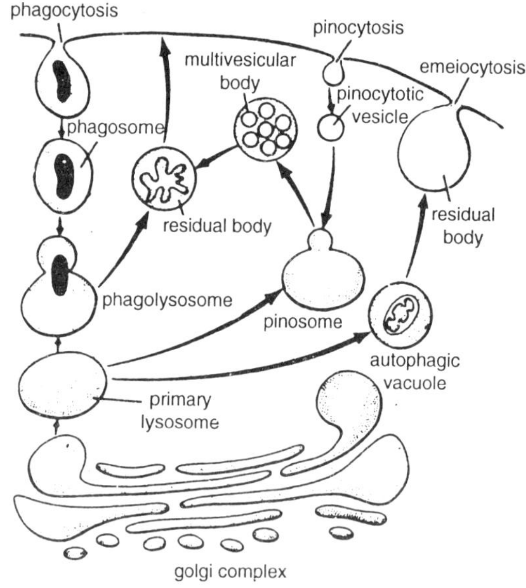


*Figure 7.3: Diagram showing biogenesis or formation of different types of lysosomes and their functioning.*

(3) *Residual or Post lysosomes.* Lysosomal membrane characterised by the presence of undigested material like myelin figures is called residual body.

(4) *Autophagic Vacuoles.* The autophagic vacuoles are also known as *autophogosomes* or *cytolysosomes.* The autopbagic vacuoles are formed when the cell feeds on its intracellular organelles such as the mitochondria and endoplasmic reticulum by the process of autophagy. In such cases, the primary lysosomes are concentrated around the intracellular organelles and digest them ultimately. The autopbagic vacuoles are formed in

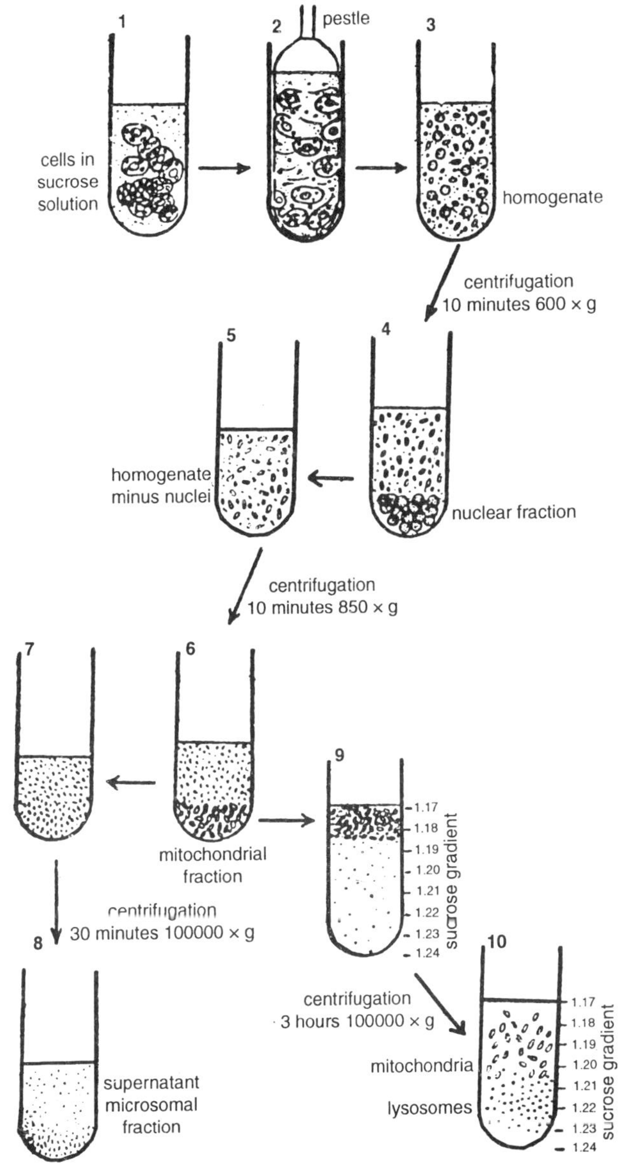


*Figure 7.4: Extraction of lysosomes by centrifugation.*

special pathological and physiological conditions. *C. de Duve* and *Allison* have observed that during starvation of the organisms many s utophagic vacuoles develope in the liver cells which feed on the cellular components.

## EXTRACTION OF LYSOSOMES

The phenomenon of centrifugation has played a great role in the study of cytoplasmic inclusion. The technique about the separation of the lysosomes has been developed in laboratory of *de Duve*. For the extraction of lysosomp, first of all the cells are homogenated in sucrose solution. Rapid mechanical rotation of the pestle ruptures the cells, setting the intracellular particles free in the medium. Then successive centrifugation of the resulting homogenate produces fractions, which can be separated by micro-needle. The entire process of centrifugation for lysosome can be studied.

## CHEMISTRY OF LYSOSOMES

Lysosomes contain variety of enzymes, upto the present time about 40 enzymes have been isolated in variety of tissue type. Some common enzymes, are β-galactosidase, β-glucuronidase, β-N-acetyl-glucosaminidase, α-glucosidase, α-mannosidase; cathepsin A (Acid protease), cathepsin B (Acid protease), aryl sulphatase A, aryl sulphatase B, acid ribonuclease, acid deoxyribonuclease, acid phosphatase, acid lipase, phospholipase A, phosphotidic acid phosphatase byaluronidase, phosphoprotein phosphatase, amino peptidase A, dextranase, saccbarase, lysozyme (muramidase), $Mg^{++}$ activated ATPase, indoxy, lacetate, esterase and plasminogen activator. All these enzymes of the lysosomoes are enclosed within the single lipoprotein membrane. Most of these enzymes function more efficiently under slightly acidic medium, pH optima around 5.0, as such they are collectively called as acid bydrolases.

**Table 7.2: Selected lysosomal enzymes.**

| Enzyme | Natural Substrates |
|---|---|
| *Phosphatases* | |
| Acid phosphataseP | hosphomonoesters |
| Acid phyrophosphatase | ATP, FAD |
| Phosphodiesterase | Phosphodiesters |
| Phosphoprotein phosphatase | Phosphoproteins |
| Phosphatidic acid phosphatase | Phosphatidic acid *Sulfatases* |
| Arylsulfatase | Aryl sulfates |
| *Proteases and Peptides* | |
| Cathepsins (several) | Proteins |

| | |
|---|---|
| Collagenase | Collagen |
| Arylamidase | Amino acid arylamides |
| Peptidase | Dipeptides *Nucleases* |
| Acid ribonuclease | RNA |
| Acid deoxyribonuclease | DNA |
| *Lipases* | |
| Triglyceride lipase | Triglycerides |
| Phospholipase | Phospholipids |
| Esterase | Fatty acid esters |
| Glucocerebrosidase | Glucocerel-.rosides |
| Galactocerebrosidase | Galactocerebrosides |
| Sphingomyelinase | Sphingomyelin |
| *Glycosidases* | |
| α-Glucosidase | Glycogen |
| β-Glucosidase | Glycoproteins |
| β-Galactosidase | Glycolipids, glycoproteins |
| α-Mannosidase | Glycoproteins |
| α—ucosidase | Glycoproteins |
| β-Xylosidase | Glycoproteins |
| α-N-Acetylhexosaminidase | Heparin |
| β-N-Acetylhexosaminidase | Glycoproteins, glycolipids |
| Sialidase | Glycoproteins, glycolipids |
| Lysozyme | Bacterial cell walls |
| Hyaluronidase | Hyaluromic acid, |
| β-Glucuronidase | chondroitin sulfate |
| | Polysaccharides, mucopoly saccharides.steroid glucuron des |

**Formation of Lysosomes**

Most of the hydrolytic enzymes present in lysosomes are glycoproteins synthesized on ribosomes bound to the endoplasmic reticulum. Like other proteins synthesized by this mechanism, insertion of the newly forming polypeptide chain through the endoplasmic reticulum is guided by a signal sequence that is subsequently removed, and oligosaccharide chains are transferred from the dolichol-phosphate carrier system to the polypeptide chain as it emerges into the lumen of the endoplasmic reticulum. The major distinction between lysosomal enzymes and other glycoproteins manufactured in this way is that mannose residues situated within the core oligosaccharide chains lysosomal enzymes become phosphorylated shortly after the oligosaccharide chain has been introduced.

The resulting *mannose-6 phosphate* groups serve to distinguish presumptive lysosomal enzymes from other glycoproteins as the various proteins synthesized within the rough endoplasmic reticulum migrate to the Golgi complex. Their recognition of lysosomal enzymes by virtue of their phosphor-ylated mannose groups is made possible by the fact that some Golgi membranes contain specific receptors that recognise phosphorylated mannose, causing the lysosomal enzymes to become selectively bound to those regions of the Golgi complex.

The unique presence of mannose-6-phosphate in lysosomal enzymes thus allows the Golgi complex to distinguish these molecules from other glycoproteins and package them into specialised vesicles destined to become lysosomes. Aftet these vesicles have budded off from the Golgi complex, the internal pH of the vesicles is lowered by the action of an active transport system that pumps protons into the interior of the vesicle.

As the pH within the lysosome is lowered, the hydrolytic enzymes are released from their binding sites on the membrane because the interaction between the phosphorylated mannose groups and their membrane receptors is disrupted at low pH. After release into the lysosome interior, the free hydrolytic enzymes tend to lose their phosphorylated mannose groups. At this point the stage is set for the normal functioning of the lysosome.

The realisation that all lysosomel enzymes catalyse hydrolysis reactions has naturally fostered the theory that lysosomes serve a degradative or digestive role within the cell. As we now see, there are at least four distinct ways in which this digestive function is utilised by cell. These include (1) degradation of foreign matter taken up by endocytosis, (2) destruction of worn-out organelles *(autophagy),* (3) breakdown of cellular structures associated with cell death *(autolysis),* and (4) digestion of extracellular materials.

**Endocytosis**

One of the most important functions of the lysosomes is its ability to catalyse the breakdown of complex macromolecules brought into the cell by the process of endocytosis. Endocytosis refers to the uptake of extracellular materials trapped in membrane vesicles that pinch off from the plasma membrane. When the materials taken up by such vesicles consist of large particles (i.e., those visible by light microscopy), the term *phagocytosis* is applied. *Pinocytosis* refers to the uptake of all other matter, including small particles and watersoluble macro molecules such as antibodies, enzymes, hormones, and toxins.

Endocytosis usually involves an interaction between the substance being ingested and plasma membrane binding sites. Some of the earliest evidence for the importance of such binding was provided by Ralph Steinman and Zanvil Cohn, who investigated the uptake of soluble and insoluble forms of the protein horse-radish peroxidase. These studies were carried out using specialised scavenger cells, termed *macrophages,* which are highly active in carrying out phagocytosis.

## DIFFERENT NOMENCLATURE

The cytochemical defination of lysosomes based on the presence of a single unit membrane and a positive staining reaction for acid phosphatase and some related enzymes, may be considered for most practical purposes equivalent to the biochemical definition. As our knowledge of the significance of lysosomes in cell physiology has progressed, it has become evident that the term lysosome covers a variety of different forms which can be distinguished on the basis of morphological and functional criteria. The following are the terms commonly used in the literature

(*a*) *Autophagic Vacuoles.* It is a membrane lined vacuole containing morphologically recognisable cytoplasmic components.

(*b*) *Cytolysome.* Same as Autophagic vacuoles.

(*c*) *Cytosome.* Particles referred to cytosomes are usually lysosomes. Some workers include the un-related micro-bodies under this term.

(*d*) *Cytosegresome.* Same as Autophagic vacuoles.

(*e*) *Micro body. A* particle found in liver and kidney, bounded by a single unit membrane and containing a finely ,granulated material. According to *de Duve,* they are definitely not lysosomes.

(*f*) *Multivesicular Bodies.* Structures lined by a single membrane and containing inner vesicles resembling Golgi complex and are considered to be lysosomes.

(*g*) *Resfdual bodies.* Membrane lined inclusions characterised by undigested residues comprises telo-lysosomes and hypothetical post-lysosomes.

# FUNCTIONS

## Lysosomal Digestion of External Particles

Large molecules are taken into the cell by the process called *phagocytosis. This is* a perfectly adequate and accurate term that implies a cell that eats. But recently the new term *endocytosis* has won favour. The first clear indication of a relationship between lysosomes and engulfment of extracellar material was provided by *Stians*. As shown in the figure, the cell engulfs the particles and then forms and invagination that becomes pinched off from the cell membrane to become an internal sac or body.

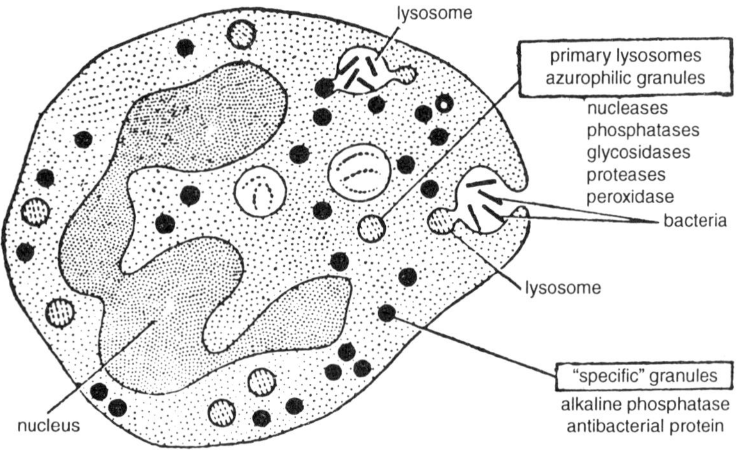


*Figure 7.5: A professional phagocyte.*

It is referred to as *a phagosome*. A phagosome then moves towards the lysosome. Exposure of the material to the lysosomal hydrolases occurs through fusion of the phagosome with a lysosome. This results in the formation of a *secondary lysosome* or *digestive vacuole*. The lysosome involved in the process may be primary or secondary, depending on the relative sizes of the two partners.

The process may appear to an observer as a lysosome discharging enzymes into a phagosome or as a phagosome shedding its centents into a lysosome, as may be the case in hepatic parenchymal cells; or simply as a mutual sharing of the contents of the two vacuoles, if they are of comparable sizes. Now the enzymes from the lysosome can come into contact with the molecules brought into the cell in the phagosome and digestion occurs. Once the molecules are digested, the digested products

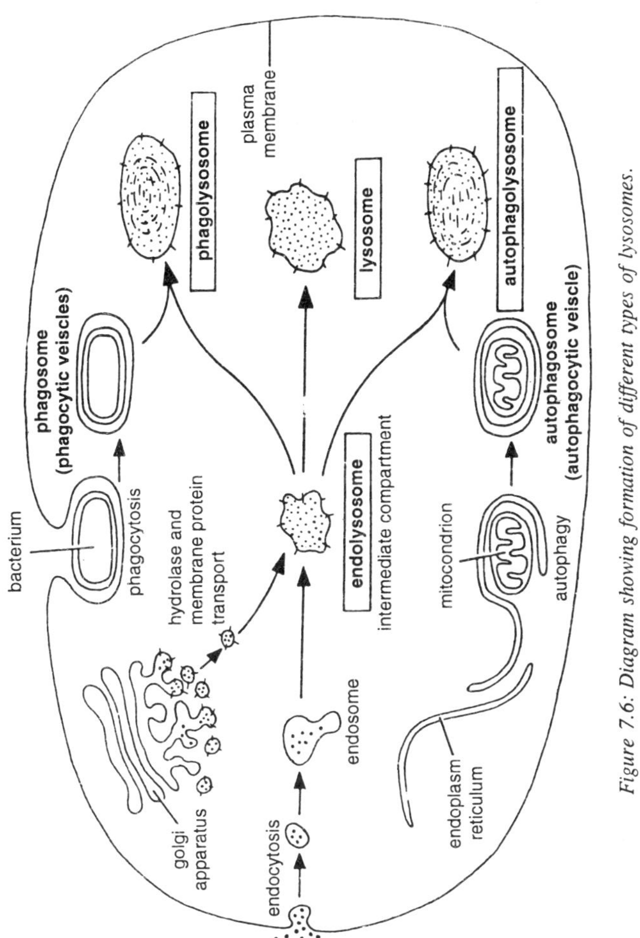


*Figure 7.6: Diagram showing formation of different types of lysosomes.*

can diffuse out of the so called digestive vacuole into the cytoplasm of the cell leaving the residue in the digestive vacuole. The digestive vacuole now moves on to the cell membrane where the so called reverse phagocytosis or defecation occurs.

### Digestion of Intracellular Substance

In certain cases portions of the cell, somehow, find their way inside the cell's own lysosomes and are broken down. This process is termed as *cellular autophagy. How* do they get in, is not clear, and the role which autophagy plays in cell function can only be summarised. Proteins, fats and polysaccharides can all be synthesized, and stored in the cell. During the starvation of cell these stored food materials are digested by lysosomes to give energy. What stimulates autophagy to take place and how do the large molecules get into the lysosome is not clear.

**Cellular Digestion**

When a cell dies, the lysosomal membrane ruptures. The liberated enzymes become free in the cell, which then quickly digest the entire cell. The hypothesis has been advanced that this is a built in mechanism for removing dead cells. In multicellular animals, many cells are constantly being formed, live for a short period of time, and then die. The self digestion may occur as a pathological mechanism just for example, if a cell is cut off from its oxygen supply or poisoned, the lysosomal membrane may rupture, thereby permitting the enzymes to dissolve the cell. Therefore, they are also considered as *suicide bags* of the cells.

**Extracellular Digestion**

A cell can discharge lysosomal enzymes to destroy the surrounding structures. This function is performed by reverse phagocytosis. A pocket of enzymes from a lysosome is released outside the cell where it then digests contigious structures. This is thought to explain, how sperms penetrate teh protective coating of the ovum during fertilisation. It may also explain how oesteoblast cells that destroyu bones, function. This may also be the explanation for the well known ability of white blood cells to pass quickly out of the blood vessels and into the tissues spaces at the site of an infection.

**Role in Secretion**

In recent years evidences have started accumulating suggesting the role of lysosomes in the formation of secretory products in secretory cells. The phenomenon of lysosomes-mediated thyroid hormones secretion is the best known example of direct lysosomes involvement in secretory process. Lysosomes also play possible role in the regulation of hormone secretion. It is believed that mammotrophic hormones of the anterior pituitary is synthesized on the ribosomes of the RER and is packaged into secretory granules by passage through the Golgi.

The epithelial cells of the thyroid also contain lysosomes rich in lysosomal enzymes. The follicles of the thyroid gland contain high molecular weight protein thyroglobulin, which is stored as colloid in the lumen. The thyroid hormones thyoxine and tri-iodothyroxin are linked with this protein. The colloid containing thyroglobulin enters the epithelial cell by pinocytosis. The colloid droplets fuse with primary lysosomes to form secondary lysosomes or digestive vaculoes. The thyroid hormones are split from the thyroglobulin and released into the blood strem. Thus the thyroid hormones are released by hydrolysis of thyroglobulin.

**Chromosome Breaks**

Lysosomes contain the enzyme deoxyribonuclease (DNAse). This enzyme causes chromosomal breaks and their rearrangement. DNAase has two active sites and breaks down both the strands of DNA. The breaks have been produced experimentally in isolated chromosomes inclubated in DNAase. These breaks lead to various syndromes.

**Role in Development and Metamorphosis**

Lysosomes are important in development. Good evidences lave accumulated on the role of lysosomes in involution of iterus and mammary glands immediately in postpartum. )urina metamornhosis. the nrocess of resorDtion of the tadpole tail and regression of the various larval tissues, including the fat body and the salivary gland, are accompained by increased lysosomal acid hydrolase activity (Weber).

**Osteogenesis**

During conversion of cartilage into bone, the special osteoblast cells produce lytic substances which erode the matrix of the cartilage and help in the formation of bone.

**Role of Lysosomes during Cell Division**

During the cell division, the lysosomes of that particular dividing cell move towards the periphery instead of near the nucleus, as in usual cases they are seen. During the cytokinesis roughly equal number of them move towards opposite poles. Sometimes during cell-division certain repressors in cytoplasm inhibit cell-division. Lysosomes secrete certain lepressors which destroy the repressor and results in cell division.

**Help in protein synthesis**

*Novikoff* and *Essner* have suggested the possible role )f lysosomes in protein synthesis. Recently, *Singh* has correlated lysosomal activity with the protein synthesis. In the liver and pancreas of some birds, lysosomes seem to be more fictive and developed showing possible relationship with cell metabolism.

**Lysosomes and Cancer**

Malignant cells are found to contain abnormal chromosomes. It is presumed that the chromosomal abnormality is caused by chromosomal breakage presumably produced by the lysosomal enzymes. The partial

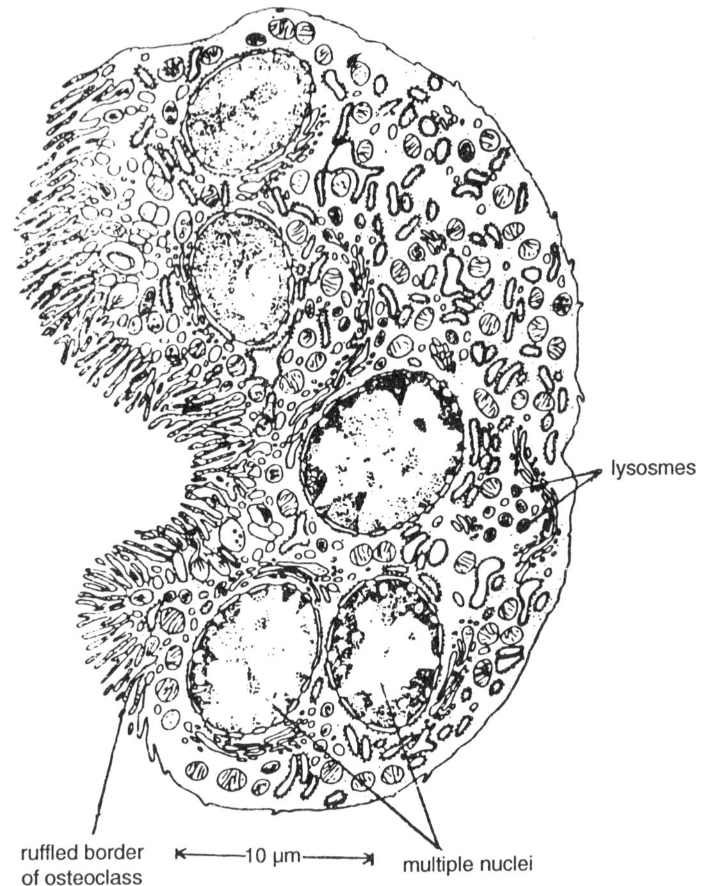


*Figure 7.7: Lysosome in an Osteoclast.*

deletion of chromosome 21 in man is associated with the chronic myeloid leukemia (blood cancer).

**Removal of Dead Cells**

*Hirsch* and *Cohn* suggested that lysosomes help in the removal of dead cells in tissue. The lysosomal membrane ruptures in these cells, releasing the enzyme into body of cell, so that whole cell may be digested. This process of tissue. degeneration (necrosis) is due to this lysosomal activity.

**Fertilisation**

During fertilisation the sperm releases hydrolytic enzymes 'rom the acrosome vesicle. These enzymes help in the penetraion of the sperm through the envelopes of the egg. Fluores;ence microscopy studies of acridine-stained spermatozoa of he guinea pig show that the acrosome vesicles contain several enzymes, including byaluronidase and proteases, which are also pound in lysosomes. In fact the acrosome vesicle has been ooked upon as *a giant lysosome*. The acrosome vesicle enzymes also apparently activate the egg by breaking down its cortical granules.

## ORIGIN OF LYSOSOMES

They have multiple origin depending upon the tissue in which they are located or on their function in a specific cell.

### Extracellular Origin

Lysosomes may be the vacuoles absorbed into cell by the process, pinocytosis. The pinocytic vacuole may later become cytoplasmic particle and thereon enzymatic activity becomes developed.

### Origin from Golgi Complex

There are evidences that lysosomes orginate from the Golgicomplex and represent zymogen granules. Their similar function and structure with Golgi complex support this view. Recent studies have shown that accumulation of secretroy products within Golgi vacuoles leads to the formation of lysosomes, and membranes surrounding the products are derived from Golgi membrane.

### Origin from ER

*Novikoff* reported that the lysosomes originate directly from granular endoplasmic reticulum, by a process of blebbing.

### Extracellular Digestion by Hydrolytic Enzymes

In addition to digesting materials contained within the cell, hydrolytic enzymes may in some instances be secreted into the extracelluler space. During fertilisation, for example, penetration of the sperm head through the outer layers surrounding the egg surface is aided by the release of hydrolytic enzymes derived from the Golgi complex of the sperm cell. Situations also occur, however, where release of hydrolytic enzymes from the cell has detrimental effects.

A case in point is the lysosomal enzyme release that often follows cell damage caused by physical trauma or microbial infection. In such instances the released enzymes may trigger additional tissue damage and inflammation. The idea that release of lysosomal nzymes can contribute to tissue inflammation is supported by ae discovery that an arthritislike disease can be produced in abbits by injecting drugs known to disrupt lysosomes. This nding explains why drugs that inhibit lysosomal enzyme -lease, such as the steroid hormones cortisone and hydroortisone, are effective antiinflammatory agents. Vitamin A, on the other hand, increases the ]ability of the lysosomal memrane and promotes discharge of hydrolases from the cell. This explains why connective tissue damage and spontaneous one fractures often occur in individuals consuming excess uantities of vitamin A.

**Lysosomal Storage Diseases**

Several dozen genetic diseases afflicting young children are .ow known to be caused by an excessive intracellular accumuition of polysaccharides or lipids. The quantity of polysacharide or lipid stored is often massive enough to interfere with and even destroy the cells involved. Depending on the particular cell types affected symptoms such as muscle weakess, skeletal deformities, and mental retardation may result.

The first of these so-called *lysosomal storage diseases* to have is underlying mechanism unraveled was *type II glycogenosis*, an illness whose victims die at an early age with abnormally arge amounts of glycogen in the liver, heart, and muscles. In 963 H. Hers discovered that type II glycogenosis is caused by a severe deficiency of the lysosomal enzyme P-glucosidase, vhich cataiyses hydrolysis of glycogen to oligosaccharides and glucose. In the absence of this enzyme, undigested glycogen Lccumulates within lysosomes.

Not only did this discovery provide an explanation for glycogen storage disease, but it also led Hers to postulate the existence of a wide spectrum of other diseases corresponding to genetic defects in particular lysosomal enzymes. He predicted that in each case, the disease symptoms are caused by an abnormal accumulation of the undigested substrates of the defective enzyme. This unifying theory has led to an understanding of the causes of what were once a bewildering array of mysterious diseases.

Although in some instances excessive storage of simple polysacch-arides or mucopolysaccharides is involved most of these storage diseases result from an abnormal accumulation of glycolipids. Because glycolipids are highly concentrated in brain tissue and are important constituents of the myelin sheath surrounding the axon, the symptoms of these storage diseases usually include severe mental retardation. The glycolipid accumulated in brain cell lysosomes can often be recognised morpholo-gically by its tendency to form unusual layered structures known as *zebra bodies.*

The discovery of the molecular basis of the lysosomal storage diseases has led to the suggestion that these illnesses might be alleviated by replacing the defective enzymes. Direct administration of missing lysosomal enzymes, however, is impractical for several reasons. To begin with, destruction of the administered enzymes by serum proteases or by the individual's immune system would minimise the effectiveness of treatment. Furthermore, many cells do not take up foreign molecules efficiently.

An alternative approach is to encaptulate the required enzyme in artificial lipid vesicles (liposomes). This encapsulation process protects the enzymes from destruction and has the added advantage that liposomes are known to be actively taken up by cells and fused with lysosomes. In this way a missing lysosomal enzyme might be delivered directly to its appropriate site of action. Encouraging results from animal studies suggest that this approach may ultimately be useful for the treatment of inherited enzyme deficiencies in humans. An alternative therapeutic approach is to attempt to correct or replace the gene coding for the defective lysoso mal enzyme.

**Killing and Digestion**

Simple enzymatic digestion by hydrolases is not the only process that occurs in secondary lysosomes. Many microorganisms, while they are viable, resist attack by hydrolases. Hence, mechanisms have evolved specifically to kill ingested organisms prior to their breakdown.

Lysosomes typically contain *lysozyme,* an en degrades bacterial cell walls. esnecially the simple gram-positive bacteria. The lysosomes of neutrophils also contain lactoferrin, which binds iron and other metal ions tightly, thus inhibiting growth of the ingested organisms. The contents of the digestive vacuoles are usually very acidic (less than pH 4 in neutrophils) which by itself inhibits growth of many bacteria. And finally, oxidases, thought to be present at the surface of the cell membrane and thus also on the inside surface of digestion vacuoles produce hydrogen peroxide from molecular oxygen. Hydrogen peroxide is microbicidal in its own right, but in neutrophils and most macrophages it is also utilised by an enzyme called myeloperoxidase to covalently fix $Cl^-$ or $I^-$ onto organic targets such as bacteria. This halogenation greatly facilitates killing of the organism.

**Autophagy**

Although lysosomes are present in probably all nucleated cells of both animals and plants, we have described their activities only in the major phagocytic cells. Their function in other cells is much the same. That is, they fuse with endocytic (including pinocytic) vesicles to destroy the vesicle and its contents. Lysosomes are involved not only in this process of digesting extracellular materials, known as *heterophagy,* but also in the digestion of intracellular substances, an activity known as *autophagy.*

In plants, for example, the cytoplasm frequently has large uoles containing reserve food supplies in a stored form. These storage Dolvmers can be broken down to useful size by fusion of the vacuole with lysosomes.

Lysosomes are similarly responsible for turnover of other orgenelles in both plants and animals, a process that is accelerated during involutional states— in a muscle that is not being used, during starvation, after injury, or druring the remodeling that occurs with cellular differentiation. Even in unstressed normal cells, the turnover of organelles by autophagy is relatily rapid. The half life of liver cell mitochondria, for ample, is five to six days, for peroxisomes it is one to two days, and for ribosomes about five days.

The autophagy of cytoplasmic organelles starts with an *isolation envelope,* which is a sack of smooth-surfaced membrane that wraps around the material to be removed and separates it into a sort of phagocytic vacuole known as an *isolation body.* The source of the isolation membrane at least in damaged liver and kidney, seems to be the endoplasmic reticulum. However, recent work by Michael Locke and his associates at the University of Western Ontario indicates that the isolation membrane in insect cells during metamorphosis, and in some other animal cells, is derived instead from the Golgi apparatus. In any case, lysosomes can fuse with the isolation body to create an autophagic digestive vacuole and release breakdown products to the cytoplasm to be used as an emergency source of food or as raw material for cellular remodeling.

### Residual Bodies

Although killing and breakdown in the lysosomes is efficient, some materials are simply not digestible. These substances may be released from the cell by fusion of the digestive vacuole with the cell membrane (exocytosis). This process is very efficient in protozoa, but not so efficient in higher animals. Hence, inactive lysosomes containing debris may accumulate.

These debris-filled organelles, called *residual bodies,* are of no consequence to cells like macrophages or neutrophils, because the cells don't live long enough for significant amounts to accumulate. In cells that do not divide, and are not replaced, substantial quantities of this material may accumulate. In neurons and muscle cells, especially, there is a steady increase in the number of these bodies, which are called in this context *lipofuscin granules*. The age of an animal can often be estimated from their concentration, but whether enough ever accumulates in a normal lifespan to compromise the function of the cell is unclear.

### Spherosomes and Related Vacuoles in Plants

The isolation of lysosomes from plant tissues has been a difficult undertaking because methods developed for animal tissues cannot be directly applied to plant lysosome isolation.

Using modified procedures, however, it is now possible to recover acid hydrolase activity in particulate fractions of plant cells. The purified fractions containing these hydrolases are enriched in small, membrane-enclosed vesicles. These structures resemble *spherosomes,* which are highly refractive spherical particles, approximately 0.5-1.0 $\mu$m in diameter, originally identified by light microscopists in the plant cell cytoplasm. Cytochemical staining for acid phosphatase in tissue sections has confirmed that spherosomes, as well as larger cytoplastic vacuoles, contain lysosomal enzymes.

A principal feature distinguishing animal lysosomes from plant spherosomes is that the latter stain intensely with fatsoluble dyes, indicating a high lipid content. The accumulation of lipid in vesicles containing hydrolytic enzymes suggests that the spherosome functions in storing and mobilising reserve lipid. In addition to its role in lipid metabolism, the spherosome and other hydrolase-containing plant vacuoles are thought to be involved in digesting and recycling intracellular constituents in a manner analogous to the animal cell lysosome. Digestion of foreign particulate matter is probably not a major function of the spherosome, however, because the presence of the plant cell wall generally prevents phagocytosis from bringing in foreign particulate matter.

## PEROXISOMES, GLYOXYSOMES AND OTHER MICROBODIES

Among the enzymes monitored by de Duve in his early idles on lysosomes was the purine-catabolising enzyme, *urate idase*. In spite of its high concentration in lysosomal fracns isolated by differential centrifugation, some minor diffeices in its distribution pattern relative to that of other osomal enzymes suggested that it might be localised in some ier organelle. When isodensity gradient centrifugation was iployed to enhance the resolution obtained, it became evint that urate oxidase is contained in a particle whose density lers from that of lysosomes. Shortly thereafter the enzymes *amino acid oxidase* and *catalase*, involved in the formation d breakdown of hydrogen peroxide, respectively, were found to behave the same way as urate oxidase. Because of the involvement of these latter enzymes with hydrogen peroxide metabolism, the organelle containing them came to be called the *peroxisome*.

The conclusion that peroxisomes and lysosomes are distinct organelles received subsequent support from experiments in which the rate of release of acid phosphatase and catalase from detergent-treated subcellular fractions was compared. If these two enzymes were located in the same particle, one would expect them to be released simultaneously as one adds increasing concentrations of detergent to rupture the vesicles in which they are contained. In fact ten times as much detergent is required to liberate catalase as compared to acid phosphatase, indicating that these two enzymes are localised in particles with differing properties.

Because of the relatively small difference in density between peroxisomes and lysosomes, isodensity centrifugation does not normally achieve a complete separation of these particles from each other. Better separation can be accomplished, however, by isolating subcellular fractions from animals that have been first injected with the detergent Triton WR 1339. This detergent is selectively accumulated within lysosomes and causes their density to decrease dramatically making possible the large-scale isolation of peroxisomes for structural and functional studies.

Electron microscopic examination of isolated subcellular fractions enriched in peroxisomes has confirmed the unique identity of these organelles. The small, roughly spherical vesicles present in such fractions resemble structures seen in tissue sections by early electron microscopists and given the general name *microbodies*. Microbodies vary from 0.2 to 2 $\mu$m in diameter and consist of a finely granular matrix surrounded by a single limiting membrane. These structures often contain elaborate crystalloid cores containing the enzyme urate oxidase. Microbodies in which such cores are present can clearly be identified as peroxisomes, but in cases where distinctive cores are not present, peroxisomes may be difficult to distinguish from lysosomes and other cytoplasmic vesicles, Unequivocal identification can be made, however, by using the cytochemical staining reaction for catalase. This reaction, based on the ability of catalase to oxidise *diaminobenzidine* (*DAB*) to an electron-dense reaction product, permits the unequivocal identification of peroxisomes. It should be emphasized that the term *microbody* is only a morphological label employed by electron microscopists when they see a "small body." Unless these structures have a urate oxidase core or are stained for catalase, they cannot be clearly stated to be peroxisomes. As shall be seen later, other types of microbodies exist in addition to peroxisomes.

Peroxisomes occur primarily in the liver and kidney cells of vertebrates, in the leaves and seeds of plants, and in eukaryotic microorganisms such as yeast, protozoa, and fungi. The spectrum of enzymes present in peroxisomes varies considerably among tissue sources. The only enzyme common to all peroxisomes is catalase which accounts for up to 15 percent of its protein. In addition to catalase, hydrogen-peroxide producing oxidases such as urate oxidase, amino acid oxidase, and glycolate oxidase are usually present. These enzymes employ molecular oxygen as oxidising agent and generate hydrogen peroxide ($H_2O_2$) according to the following general reaction

$$RH_2 + O_2 \rightarrow R + H_2O_2$$

**Table 7.3 : Major enzyme activities of peroxisomes.**

| | *Source of Peroxisomes* | | |
|---|---|---|---|
| | | *Plant* | |
| **Enzymatic Activity** | **Rat Liver** | **Seedlink** | **Plant Leaf** |
| 1. Catalase | + | + | + |
| 2. Urate oxidase | + | + | + |
| 3. D-Amino acid oxidase | + | + | – |
| 4. L-Ammo acid oxidase | + | + | – |
| 5. L-$_a$- Hydroxy acid oxidase | – | + | + |
| 6. Glycolate oxidase | + | + | + |
| 7. Fatty acid a-oxidation | + | + | – |
| 8. Citrate synthase | – | + | – |
| 9. Aconitase | | + | – |
| 10. Isocitrate lyase | – | + | – |
| 11. Malate synthase | – | + | |
| 12. Malatc dchydro2enase | – | + | + |
| 13. Glyoxylate reductase | + | – | + |
| 14. Glyoxvlate transaminase | – | – | + |

The $H_2O_2$ formed by reactions of this type can subsequently be hydrolysed by catalase acting in one of two alternative ways. In the so-called *catalatic mode,* one molecule of $H_2O_2$ donates its electrons to another molecule of $H_2O_2$, forming a molecule of water and a molecule of oxygen catalase

$$H_2O_2 + H_2O_2 \rightarrow O_2 + 2H_2O$$

In the *peroxidatic mode* the electron donor is a substrate other than $H_2O_2$:

$$RH_2 + H_2O_2 \rightarrow R + 2H_2O$$

In either case, the net result is the breakdown of hydrogen peroxide.

The hallmark of peroxisomes is therefore the association of $H_2O_2$-producing oxidases with the $H_2O_2$-degrading enzyme, catalase. Although the exact functional significance of this association is not certain, some of the more likely possibilities will be discussed below.

**Metabolic Functions of Peroxisomes**

Because hydrogen peroxide is toxic, it might seem logical to conclude that the peroxisome functions to protect cells from exposure to this substance by compartmentalising the enzymes that produce and destroy it. The flaw in this argument, however, is that many H2O,-producing oxidases occur in the cell sap and inmitochondria, suggesting that the peroxisome contributes little to the protection of the cell from the effects of $H_2O_2$. This conclusion is further reinforced by the fact that many cells producing $H_2O_2$ do not even contain peroxisomes. It therefore appears that the need for peroxisomes must be related to the metabolic pathways these organelles carry out rather than with the simple need to protect cells from $H_2O_2$. Examination of the various enzymes present in peroxisomes has led to consideration of the following possibilities.

1. *Inactivation of toxic substances.* One possible role of peroxisomes may involve the coupling of $H_2O_2$ degradation to the inactivation of toxic compounds. Ctalase is the most active enzyme present in the peroxisome, and therefore legrades hydrogen peroxide at a much faster rate than it is ormed by other peroxisomal reactions. The rusulting scarcity )f hydrogen peroxide favours the peroxidatic mode of catalase action, in which an alternative electron donor is substituted for :he second molecule of $H_2O_2$. The electron donors employed For this purpose include methanol, ethanol, phenols, nitrites, Formaldehyde, and formate. Oxidation of these compounds Iriven by the breakdown of $H_2O_2$ results in the detoxification of what are otherwise noxious substances.

2. *Regulation of oxygen tension.* Because peroxisomal oxidases employ molecular oxygen as an oxidising agent, the reactions catalysed by these enzymes may have a significant effect on oxygen levels within the cell. In liver cells, for example, as much as 20 percent of the total oxygen consumption is I accounted for by peroxisomes. Most of the remaining oxygen consumption occurs in mitochondria, where energy released during the oxidation of organic fuels by molecular oxygen is used to drive formation of the energy conserving molecule, ATP. Hence

*respiration,* a term referring to the oxidation of organic fuels by molecular oxygen, occurs in both mitochondria and peroxisomes. The major difference is that a portion of the energy released during respiration in mitochondria is trapped as ATP, while in peroxisomes the energy is all lost as heat.

Peroxisomal respiration also differs from mitochondrial respiration in its sensitivity to oxygen concentration. Mitochondria) respiration occurs at a maximal rate when the oxygen concentration is around 2 percent, with further increases in oxygen producing no further increase in respiratory rate. The rate of peroxisomal respiration, on the other band, increases in roughly direct proportion to oxygen tension. This disparity gives mitochondria an advantage over peroxisomes in utilising small amounts of oxygen, but also allows peroxisomes to exceed the respiratory ability of mitochondria at high oxygen concentrations. This property of peroxisomes may help to protect cells from the toxic effects of high concentrations of oxygen. Support for this idea has been obtained from studies carried out on plant cells exposed to bright light to enhance photosynthetic activity. The resulting increase in oxygen production

Although photorespiration per se does not occur in animal tissues, an analogous enhancement of peroxisomal respiration takes place when oxygen tension rises in the cell because of a reduction in the mitochondrial respiration rate. If the depression in mitochondrial respiration rate has been caused by a lack of oxidisable substrates, some of these substrates may be regenerated by the peroxisomal oxidation reactions stimulated by the rise in oxygen tension. Such a feedback system .permits regulation of mitocbondrial respiration by peroxisomal respiration, and *vice versa.*

3. *Regeneration of cytoplasmic NAD*$^+$. Many of the oxidised compounds generated by peroxisomal respiration can be converted back to their reduced forms by NADH-dependent enzymes present in the cell sap. For example, pyruvate produced in peroxisomes by the oxidation of lactate can be reduced in peroxisomes by the oxidation of lactate can be reduced back to lactate in the cell sap, accompanied by the conversion of NADH to NAD$^+$. As will be described in, this conversion of NADH to NAD$^+$ is useful in cases where NADH is being produced in the cell sap by glycolysis faster than it can be regenerated back to NAD$^+$.

4. *Metabolism of nitrogenous bases, lipids, and carbohydrates.* Because uric acid is a degradation product of the purine bases present in DNA and RNA, the localisation of urate oxidase and other enzymes of purine metabolism in peroxisomes suggests that this organelle plays a role in the breakdown of nitrogenous bases derived from nucleic acids.

Enzymes involved in the breakdown of fatty acids, and in the metabolism of various carbohydrates and organic acids, have also been identified in peroxisomes. The particular combination of enzymes present suggests the possibility that peroxisomes are involved in *gluconeogenesis,* the synthesis of carbohydrate from fats and other noncarbohydrate materials. This speculation is reinforced by the fact that the only vertebrate cell types with significant numbers of peroxisomes are liver and kidney, tissues known to be major sites of gluconeogenesis. However, the role of peroxisomes in gluconeogenesis hes been most firmly established in plant seedlings where a special type of peroxi some known as a glyoxysome is involved.

**Glyoxysomes and the Glyoxylate Cycle**

Many plants store large amounts of lipid in their seeds for subsequent use as an energy source during germination. At the appropriate time, this lipid is converted to carbohydrate by a pathway that can be divided into two major parts. The first part involves degradation of fatty acids by fl-oxidation, a process in which two-carbon fragments derived from the fatty acid chain are successsively released in the form of a molecule called acetvl-CoA. These two-carbon units are then converted to carbohydrate by the .*glyoxylate cycle,* a modified version of the Krebs tricarboxylic acid cycle. In essence, the glyoxylate cycle by-passes the $CO_2$ evolving steps of the Krebs cycle with the aid of two enzymes not present in animal cells, isocitrate lyase and malate synthase. The net result of glyoxylate cycle is the conversion of two molecules of acetyl-CoA to one molecule of succinate, a precursor for the synthesis of various carbohydrates.

In the late 1960s Harry Beevers and his associates discovered that in plant seedlings, the enzymes of the glyoxylate cycle and fatty acid β-oxidation are localised in particles that can be separated by isodensity centrifugation from both mitochondria and chloroplasts. These particles, logically named *glyoxysomes,* also contaia catalase and several $H_2O$-producing oxidases. Hence they represent a special type of peroxisome in which the complete set of glyoxylate cycle enzymes is present along with the normal $H_2O_2$-associated enzymes.

True glyoxysomes bearing the complete set of enzymes required for gluconeogenesis via the glyoxylate cycle have thus far been identified only in plant seedlings. Protozoa and yeast have peroxisomes containing some, but not all, of the glyoxylate cycle enzymes. In such cases the glyoxylate cycle may be carried out by cooperation between mitochondria and peroxisomes, each executing a portion of the cycle. In higher animals fatty acid p-oxidation has been observed in peroxisomes, but the products of this pathway cannot be used for gluconeogenesis by the glyoxylate cycle because this cycle does not occur in higher animals.

**Biogenesis of Peroxisomes**

Two theories have been proposed to explain how peroxisomes are manufactured by cells. The first model suggested that peroxisomal proteins, like secretory proteins, are synthesized on ribosomes bound to the endoplasmic reticulum these newly formed peroxisomal proteins were though to pass into the cisternae of the ER and into outpocketings that pinch off to form peroxisomes. A more recent model proposes that peroxisomal proteins are synthesized on free ribosomes, released into the cell sap, and then taken up by peroxisomes. According to this latter view, new peroxisomes are formed by expansion and budding of existing peroxisomes.

Early support for the first model was obtained from electron micrographs in which peroxisomes were seen to exhibit "tails" believed to represent connections to the endoplasmic reticulum. However, peroxisomal and ER membranes are difficult to distinguish from one another in electron micrographs, and it can be just as easily argued that the "tails" represent connections between an interconnected network of peroxisomes.

Hence in order to clearly distinguish between the alternative models of peroxisome biogenesis, its has been necessary to carry out biochemical studies designed to determine where peroxisomal enzymes are synthesized within the cell. Subcellular fraction studies have therefore been carried out on tissues briefly incubated in the presence of radioactive amino acids. According to the first model, the newly formed radioactive peroxisomal proteins should first appear associated with the rough ER of the microsomal fraction. The experimental data have failed to support this prediction, however. Instead, peroxisomal proteins such as catalase first appear in the cytosol fraction and only later become concentrated in peroxisomes, suggesting that they are being synthesized on free ribosomes.

Additional support for this conclusion has come from the demonstration that peroxisomal enzymes are synthesized by messenger RNAs isolated from free, but not membrane-bound, ribosomes. The notion that peroxisomal proteins are not synthesized like secretory proteins is further supported by the discovery that the major peroxisomal proteins are neither glycosylated nor synthesized as precursors containing a signal sequence.

The above findings raise the significant question of how peroxisomal proteins made in the cell sap are selectively transported into peroxisomes. Although the answer to this question is not completely understood, it is though that specific proteins localised within the peroxisomal membrane aid in the recognition and uptake process.

**Evolutionary Origin of Peroxisomes**

Although several possible functions of the peroxisome have been mentioned, there are reasons for questioning whether any of these are important enough to justify the existence of such an organelle, especially in the cells of higher animals. To begin with, most vertebrate cells get by perfectly well without peroxisomes, and even when they are present their necessity is questionable. Humans inheriting genetic deficiencies of the enzyme catalase, for example, exhibit no obvious. disease symptoms. The peroxisomal enzyme urate oxidase is absent in many organisms, including humans, and the function of the peroxisomal enzyme that oxidises D-amino acids is somewhat puzzling because it is the L-amino acids that are normally found in protein molecules. Finally, most of the metabolic events supposedly occurring in peroxisomes can take place elsewhere in the cell.

The above considerations have spawned the speculation that the present-day peroxisome is a fossil organelle derived from an ancestral particle that performed critical functions hundreds of millions of years ago, but is no longer needed and is in the process of dying out. According to this theory, the original peroxisome formed when oxygen first appeared on earth. Oxygen can be toxic because it tends to react with 'biological molecules to form hydrogen peroxide. Adaptation to the appearance of oxygen in the atmosphere therefore required the evolution of a system for disposing of hydrogen peroxide.

The peroxisome was especially well suited to this task because it uses the oxygen-induced production of hydrogen peroxide to facilitate the oxidation of intermediates involvbd in cellular carbohydrate metabolism. The major disadvantage of peroxisomal respiration, however, is that the energy released during this carbohydrate oxidation is not - conserved in a useful chemical form, but is dissipated as heat. When mitochondria appeared later in evolution, their ability to link respiration to the formation of ATP permitted them to prevail over peroxisomes. Hence the peroxisome gradually lost many of its enzymes and became less and loss important.

Some biologists believe that mitochondria evolved from :ancient bacteria that were engulfed by primitive nonbacterial -cells hundreds of

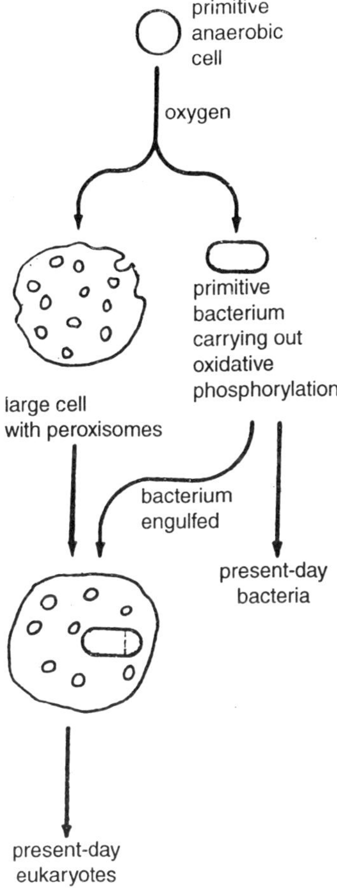


*Figure 7.8: A hypothetical model proposed to explain the evolutionary origins of peroxisorrres and mitochondria.*

millions of years ago. Such a notion can be combined with our picture of the ancestral peroxisome to form an integrated theory of the evolution of respiration. This theory asserts that when oxygen first appeared on the earth, cellular life diverged in two directions. One resulted in relatively small cells that linked respiration to ATP formation, while the other produced larger cells that depended on peroxisomes for respiration. At a later time some of the smalier, bacteria-like cells were engulfed by the larger, peroxisome-containing cells. The engulfed bacteria ultimately evolved into mitochondria, displacing the peroxisomes in importance.

In spite of the apparent superiority of mitochondria in respiration, one should not overlook the fact that peroxisomes have persisted in the presence of mitochondria for hundreds of millions of years, albeit only in certain cell types. This persistence must reflect some useful function, though we are not certain what it may be. In plant cells the glycoxylate cycle certainly falls in this category, but the enzymes of this cycle were lost during the course of animal evolution, and are e-en lacking in the leaves of green plants. Thus the reason for the evolutionary persistence of peroxisomes in most cell types remains somewhat of a mystery.

# 8

# NUCLEUS

Chromosomes as the physical basis for the genetic properties of cells and for inheritance from parent cells to daughters have been discussed in a variety of contexts in earlier chapters. This chapter deals with the molecular structure of chromosomes and with the replication of chromosomes in preparation for cell division.

The primary property of chromosomes, namely the carrying of genes in DNA molecules, is the same in the chromosomes of prokaryotes and eukaryotes. However, prokaryotic and eukaryotic chromosomes differ in both structure and mode of replication in several fundamental ways, and it is convenient to discuss the two types of chromosomes separately. In spite of the differences, the principles of structure and replication already uncovered for prokaryotic chromosomes serve as a basis from which to analyse the much larger and structurally complex chromosomes of eukaryotes.

## PROKARYOTIC CHROMOSOMES

A bacterium contains only one chromosome, which encodes all of the genes needed for the life of the cell. Some bacteria contain a very small molecule of DNA called a *plasmid*, which encodes only a few genes. Some plasmids are valuable to the cell because they confer antibiotic resistance; however, in general, plasmids are dispensable.

### Chromosome Size

Several methods have been developed to measure either the length or molecular weight of the DNA in an intact prokaryotic chromosome. Each of these methods requires chemical extraction of unbroken DNA from cells. Extracting unbroken DNA is difficult because thin molecules easily break by mechanical scission or shear and DNases are present that degrade DNA. The DNA double helix is a stiff molecule that is easily broken by shearing forces created during stirring, pouring, or pipetting of a DNA solution. The longer the DNA molecule, the greater the probability of shearing. With simple precautions molecules with molecular weights as large as $10^8$ (150,000 bp) can be isolated without shearing. However, enzymatic breakage can be inflicted by DNases that are activated when a cell is lysed. Because of their small size the DNA molecules of plasmids are the easiest to isolate.

**Table 8.1: Sizes of DNA Molecules in Some Bacteria.**

| | *Molecular Weight* | *Number of Base Pairs* |
|---|---|---|
| *Plasmid (in E, coli)* | $5.0 \times 10^6$ | $7.50 \times 10^3$ |
| *Bacteriophage T4* | $110.0 \times 10^6$ | $0.165 \times 10^6$ |
| *Mycoplasma pneumoniae* | $0.50 \times 10^9$ | $0.75 \times 10^6$ |
| *Acholeplasma laidlawaii* | $1.0 \times 10^9$ | $1.50 \times 10^6$ |
| *Haemophilus influenzae* | $1.01 \times 10^9$ | $1.52 \times 10^6$ |
| *Neisseriaceae gonorrhoea* | $1.28 \times 10^9$ | $1.92 \times 10^6$ |
| *Staphylococcus aureus* | $1.43 \times 10^9$ | $2.15 \times 10^6$ |
| *Streptococcus faecalis* | $1.47 \times 10^9$ | 2.21 × 106 |
| *Bacillus cereus* | $2.60 \times 10^9$ | $3.90 \times 10^6$ |
| *E. coli* | $3.13 \times 10^9$ | $4.70 \times 10^6$ |
| *Salmonella puilorum* | $2.83 \times 10^9$ | $4.25 \times 10^6$ |
| *Serratia marcesans* | $5.56 \times 10^9$ | $8.34 \times 10^6$ |
| *Pseudomonas aeruginosa* | 6.96 × 109 | $10.44 \times 10^6$ |

Early estimates of the size of bacterial DNA were low because of shearing and enzyme degradation during preparation. These problems have been solved, and we now know that the sizes of DNA molecules of chromosomes in various bacterial species cover a 15-fold range from 750,000 by to a little over 10,000,000 bp. Table elsewhere in this chapter gives the sizes of DNA molecules, measured by various methods, for some bacterial species, a bacteriophage, and a plasmid. Why the DNA molecules of some species of bacteria are so much larger than others is not fully understood. Some bacteria have more genes, giving that cell greater functional versatility. However, it is unlikely that differences in gene number account for more than a part of the differences in DNA amounts.

One of the first accurate estimates of the molecular weight of

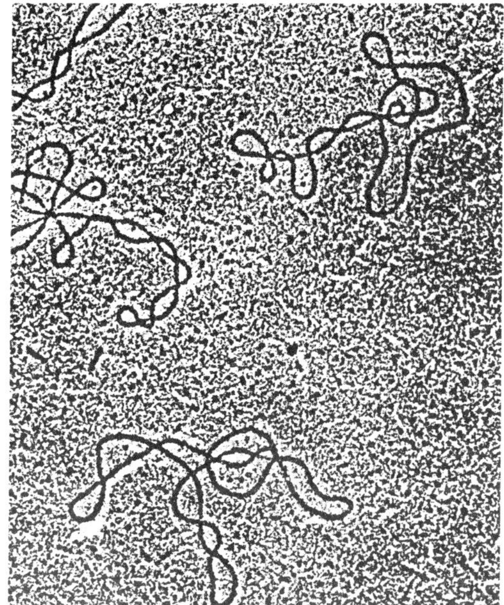

*Figure 8.1: Circular DNA molecule of a bacterial plasmid. The circles are twisted because the DNA has coiled upon itself.*

DNA in the chromosome in *E. coli* was made by autoradiography. Cells were grown for two generations with $^3$H-thymidine to label chromosomes throughout their lengths; then the DNA was extracted very carefully and spread out on a surface for autoradiography. From measurements of the length of autoradiographic images, the DNA was estimated to be at least 1100 $\mu$m long, which corresponds to 3,300,000 by or 3300 kilobase pairs (kbp). For interconversion of units of size for DNA it is only necessary to remember the 1-2-3 rule:

$$1\ \mu\text{m of DNA duplex} = 2 \times 10^6 \text{ molecular weight} = 3 \text{ kbp}$$

Three kbp have an absolute mass of $3.1 \times 10^{-6}$ picograms. One picogram (pg) equals $10^{-12}$ g. More recent measurements by several methods have shown that the chromosome of *E. coli* is about 4700 kbp.

The sizes of smaller molecules, such as those of bacteriophages and plasmids, can often be measured directly by electron microscopy. In general, however, the large molecules of bacterial chromosomes cannot be measured by electron microscopy because the field of view in an electron microscope is small compared with the length of the DNA. An exception is the DNA molecule in the wall-less bacteria, the *Mycoplasma*. These are the smallest kinds of bacteria, some measuring

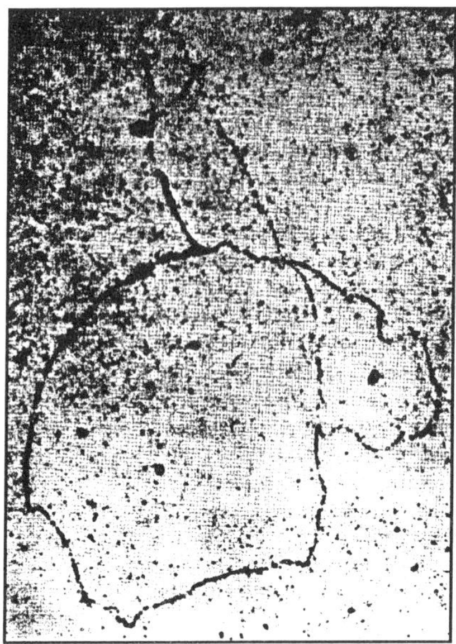

*Figure 8.2 : Autoradiograph of an intact replicating chromosome of E. coli that was labeled by incorporation of $^{3}H$-thymidine. The circular DNA molecule was undergoing replication at the time of isolation and therefore has a partially doubled structure.*

only 0.3 $\mu$m in diameter. The intact DNA molecule extracted from *Mycoplasma hominis* measured by electron microscopy is about 250 Am long, which corresponds to 750 kbp. The mycoplasma contain the smallest cell chromosomes known. Assuming as a rough estimate 1000 by per gene (coding sequence plus regulatory sequences) and assuming that all of the DNA is used for genes, the mycoplasma have an upper limit of *750* genes. About *350* kinds of protein molecules have been identified in mycoplasma. Some proteins may have escaped detection, so the number of protein-encoding genes may be somewhat higher than *350.* In addition, there are genes encoding rRNA and tRNA molecules. Therefore, the true number of genes in mycoplasma is somewhere between about *400* and *750.* Because of its relatively small size the DNA molecule in mycoplasma might be the first cellular chromosome for which the entire nucleotide sequence will be determined. From the sequence virtually every gene and its product will be identifiable.

The chromosome in *E. coli* is over five times larger than the one in mycoplasma (4700 vs 750 kbp) and could encode 4700 genes. Genetic studies indicate that the number may be 1500 to 2000. Compare this number to the 50,000 to 100,000 different genes estimated to be present in a human cell.

**Circularity of Prokaryotic Chromosomes**

The DNA molecule of all bacterial chromosomes and plasmids is a closed circle. Circularity was first suggested by genetic maps constructed for the chromosome of *E. coli*. One could explain the positional relationship among various genes in the chromosome by assuming that the chromosome was continuous, that is, was a circle. But there were other explanations for the positional relationships among genes. Proof of circularity came from physical studies of bacterial chromosomes.

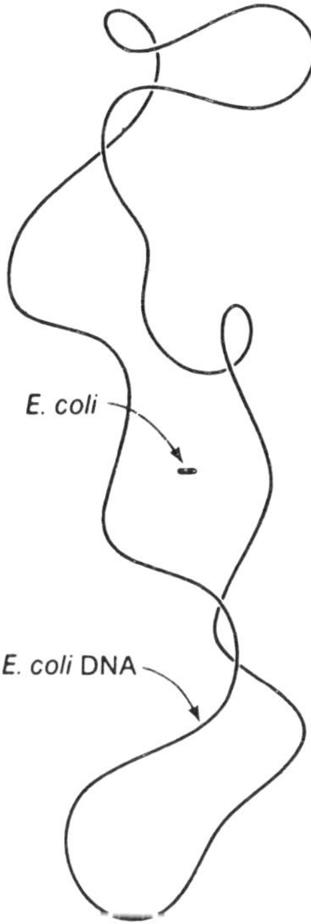


*Figure 8.3 : Schematic drawing showing the relative sizes of an E. coli cell and its DNA molecule, drawn to the same scale except for the width of the DNA molecule, which is enlarged in width approximately $10^6$ times.*

The first physical evidence of circularity was obtained from the autoradiographic study of *E. coli* chromosomes, shown earlier in Figure elsewhere in this chapter. This particular chromosome is in an intermediate stage of replication as reflected by the doubling of part of the circle. DNA molecules isolated from a variety of plasmids and prokaryotic viruses have been shown by electron microscopy to be closed circles. The chromosome of *M. hominis* (750 kbp) is the largest chromosome shown to be circular by electron microscopy. The signifi-

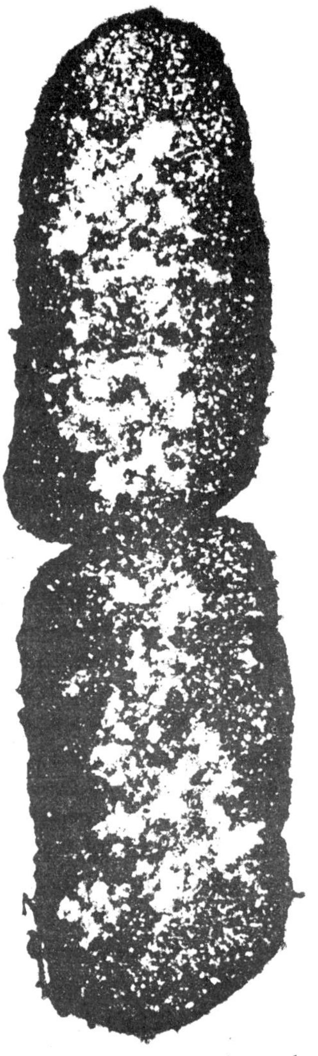

*Figure 8.4 : A dividing E. coli cell. The daughter nucleoids in each cell occupy the lightly stained areas.*

cance of circularity of chromosomes is not known, but it is important in chromosome replication (discussed later).

**Structural Organisation of Bacterial Chromosomes**

Bacterial cells measure no more than a few micrometers in their maximum dimension but contain a DNA molecule hundreds of micrometers in length. *E. coli* is a cylinder about one $\mu$m in diameter and two $\mu$m long and contains a circular DNA molecule about 1560 $\mu$m long. The DNA molecule fits into the bacterial cell by being compactly folded in a way that does not interfere with transcription or replication. The folding is obvious in an electron micrograph of a bacterium. The DNA, complexed with protein molecules, forms a dense fibrous mass, the nucleoid. Removal of the tough cell wall by digestion with the enzyme lysozyme allows the cell to be gently lysed. The nucleoid can then be separated from the cytoplasm and studied by electron microscopy.

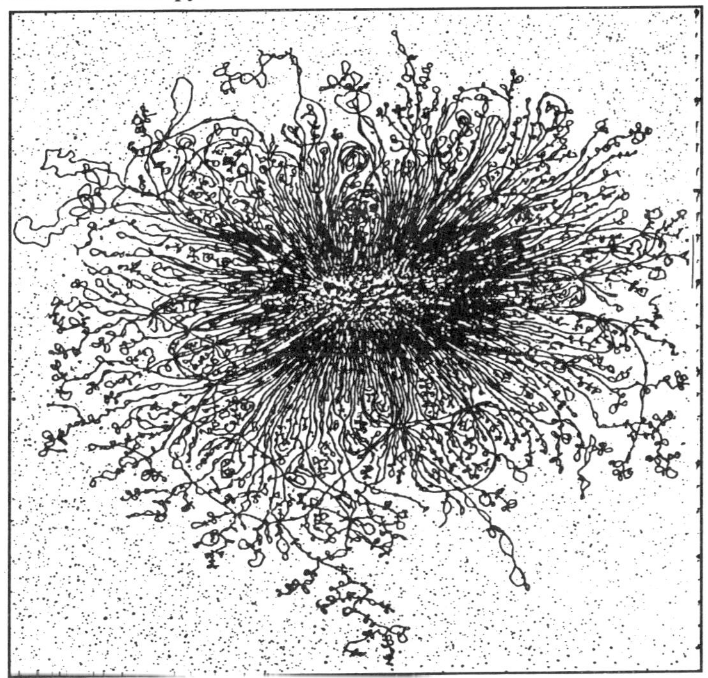

*Figure 8.5: Electron micrograph of an E. coli chromosome partially released from a cell and spread over a wide area. Many loops extend out from the central region.*

The DNA is held in the compacted state in the nucleoid by several kinds of proteins. Some of these proteins aggregate to form particles that in turn bind to DNA by means of ionic bonds between positively charged side groups of amino acids, particularly the amino group ($—NH_3^+$) of arginine, and the negatively charged phosphate groups of the DNA. The DNA coils around successive protein aggregates, producing a beaded string in which the extended DNA molecule is drawn into a shortened configuration. The beaded string is itself drawn into a more compacted state, probably by complexing with other proteins, to form the nucleoid. The beaded string, which in its fully extended state would be several hundred micrometers in length, is organised into loops called *domains*, with all the loops anchored at their two ends to one central core.

This organisation becomes clear when certain of the proteins, including those that aggregate to form the beads, are removed from an isolated nucleoid. Loops of DNA, each representing a domain containing many genes, extend out from the central protein core of the nucleoid. Because the chromosome is a continuous, circular DNA molecule, one loop must be joined to the next within the central core. This scheme of the organisation of the bacterial chromosome in a nucleoid is shown in the drawing in Figure elsewhere in this chapter.

The organisation of DNA molecules into loops or domains is thought to be important in transcription. The DNA double helix within a loop is *underwound*. To visualise what this looks like an analogy with a piece of string is useful.

Some kinds of string consist of two threads wound around each other to form a double helix although the demonstration will work whatever the number of threads. When the two ends of the string are held and then twisted in a direction intended to unwind the two threads, the string coils upon itself, that is, becomes supercoiled.

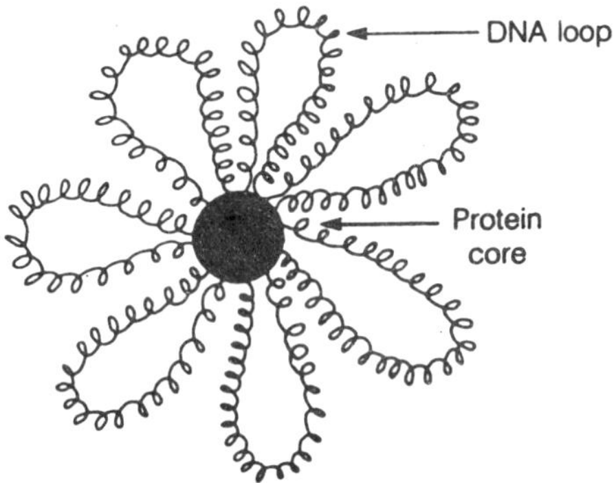


*Figure 8.6: Schematic diagram of the organisation of a bacterial nucleoid. A continuous circular molecule of DNA is organised into loops or domains that are anchored in a central core of protein. The DNA in each loop is coiled. Only seven of the approximately 50 loops are shown.*

If one end of the string is released, the string is free to rotate on its own axis and the supercoiling is released. Supercoiling induced by underwinding a double helix is called *negative supercoiling*. Supercoiling can also be produced by overwinding a double helix; such supercoiling is said to be positive. Negative supercoiling is shown for the circular duplex molecule of a bacterial virus in Figure elsewhere in this chpater. Because the molecule has no free ends the supercoiling is stable. If a break is introduced into the phosphodiester backbone of the duplex, the bond opposite the break (in the other chain) will act as a swivel and allow the supercoiling to be released.

The portion of the bacterial DNA double helix in a loop is underwound and is therefore negatively supercoiled. The DNA is prevented from rotating and releasing the supercoiling because the two ends of a loop are immobilised by attachment to proteins in the central core of the nucleoid.

Twisting of the DNA double helix to produce negative supercoiling is catalysed by an enzyme called *DNA gyrase*. The proteins that bind to DNA to form the beaded structure of the chromosome then stabilise the supercoiled configuration of individual loops.

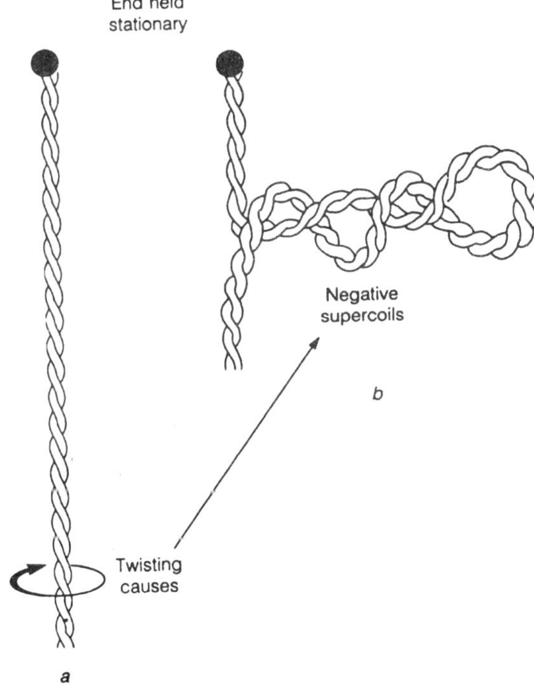


*Figure 8.7: Supercoiling in a piece of string. (a) A piece of string immobilised at the top end and twisted on its axis at the bottom end in a direction opposite to the coiling of the threads in the string. (b) Twisting causes the string to writhe upon itself, forming negative supercoils. In the same way, twisting a DNA duplex in a direction opposite to the coil of the duplex causes superhelical twisting of the duplex.*

Negative supercoiling tends to destabilise the basepair binding between the two chains in the DNA double helix, allowing the chains to separate in localised regions. This separation is believed to facilitate transcription by making the coding strand of the DNA double helix more accessible to RNA polymerase.

## Replication of Bacterial Chromosomes

Replication of the chromosome is a central event in the reproduction of a cell. During cell division the two identical DNA molecules produced by replication are distributed to the two new daughter cells. After cell division the DNA molecule in each daughter cell begins to replicate in preparation for the next cell division. What triggers DNA replication is not known but it is coordinated with the growth of the cell.

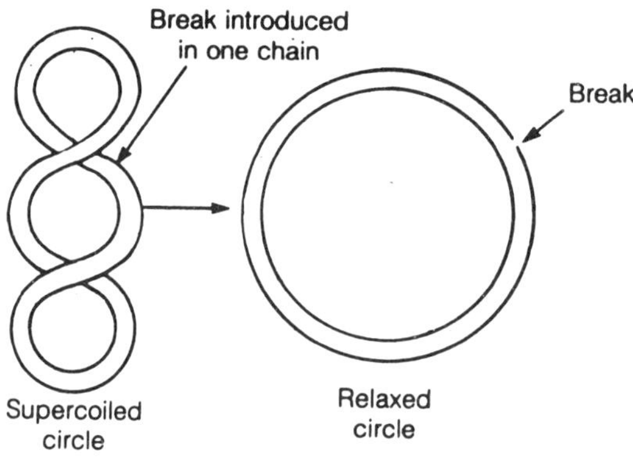


*Figure 8.8: A break introduced into one strand of a supercoiled duplex allows the duplex to rotate on its own axis and form a relaxed or open circle.*

### *Initiation of DNA replication*

Initiation of replication in *E. coli* begins at a special region of the DNA molecule called the *origin* or *ori.* The origin has been defined by mutational analysis as a 245-bp stretch of the chromosome. That is, mutations in which individual pairs are changed within the 245-bp region decrease the efficiency of initiation of replication. Base changes in adjacent regions are without effect on initiation.

*Ori* contains several short, repeated sequences that are probably recognised by proteins required for initiating DNA replication. Three of these proteins are encoded by genes designated *DnaA, DnaB,* and *DnaC.* Their roles in initiation are only partially understood. The first step in initiation is binding of the protein encoded by *DnaA* to the origin, and hence the DnaA protein has been named the *initiator protein.* What controls the timing of binding (and therefore the timing of initiation) or how it causes initiation are not known. The function of the protein encoded by the gene *DnaC is* also essential, but precisely what it does is less clear.

The gene *DnaB* encodes a protein that is required to promote unwinding of the double helix (a *helicase*). Unwinding gives a special RNA polymerase called *primase* access to the two strands as templates for replication. The RNA primase synthesizes a short RNA primer, as described in othere chapter of this book and as shown in Figure elsewhere in this chapter. DNA polymerase then takes over from RNA primase, continuing to extend the new chain with deoxynucleotides.

The unwinding of the double helix promoted by the DnaB helicase causes rotation of the double helix. This rotation is accomplished without inducing supercoiling by the breaking of one chain (nicking), which allows the bond opposite the nick in the other chain to act as a swivel that releases the torsion created by unwinding.

Swiveling is done by an enzyme called *topoisomerase*, which continuously introduces and reseals nicks ahead of replicating templates to permit progressive unwinding as replication proceeds. Another protein binds to single-stranded DNA and is called ***single-stranded binding protein*** or ***SSB***. This protein holds the two strands of DNA in the single-stranded state ahead of DNA polymerase as it moves along the single-stranded template.

Once replication has been initiated with an RNA primer, DNA polymerase works continuously on one template strand but works discontinuously on the opposite strand by the backfilling mechanism described in other chapter of this book. The first region of the chromosome to replicate is the origin, yielding origins in two partial daughter chromosomes.

Whatever condition in the cell caused the parental origin to be activated, it has dissipated, since the two new origins are not used to initiate new rounds of replication until the cell has grown, divided, and again met conditions needed for initiation. A molecular definition of the requirements for initiation of DNA replication and the molecular relation of initiation to cell growth remains one of the most important unsolved problems in cell biology.

***Bidirectional replication***

Figure elsewhere in this chpater shows the synthesis of an initiation RNA primer on each strand of the newly opened DNA double helix in the origin giving rise to bidirectional replication. By the rules of polarity of the chains in a double helix, two RNA primers are synthesized in

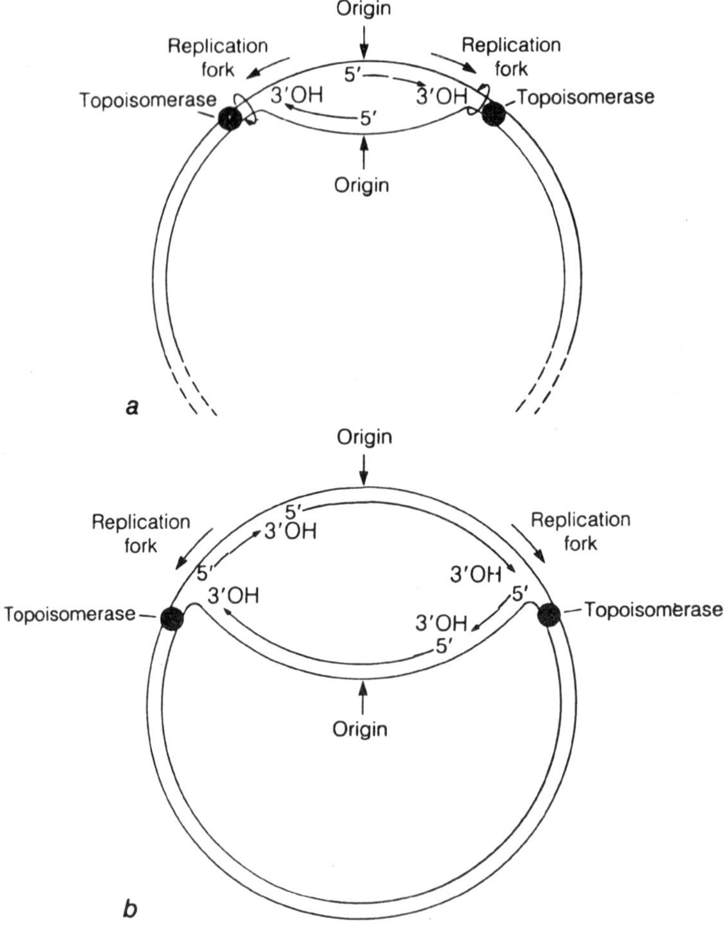


*Figure 8.9: (a) Replication begins with unwinding of the chromosome at the origin and synthesis of primer RNA on each parental strand. Topoisomerase continuously relieves torsion in the duplex ahead of each replication fork as parental DNA continues to unwind. (b) Synthesis of additional RNA primers occurs in the 5' to 3' backfilling along each parental strand. The very beginning of replication forms origins in the daughter chromosomes that are not activated until reinitiation conditions are established in the cell.*

a 5'→ 3' direction, one on each parental template, with nucleotides being added at the end with the free 3'OH group. DNA polymerase replaces primase and extends the RNA chain with deoxynucleotides in the same direction as the primate.

The continuous extension of RNA initiation primers and then DNA chains leaves a singlestranded parental chain immediately opposite. These are converted to double-helix DNA by additional RNA primers made as the parental double helix unwinds. These RNA primers are extended as DNA chains, the RNA primers are removed by exonuclease, and the individual segments of DNA are joined by DNA ligase to complete the double helix.

Because filling in replication with Okazaki fragments necessarily comes after the synthesis of the continuous chain, called the *leading strand*, the backfilled strand is called the *lagging strand*. The overall process as shown in Figures elsewhere in this chapter leads to bidirectional replication of the chromosomal DNA, with two replication forks moving in opposite directions from the origin. Autoradiography provided direct visualisation of bidirectional replication in the following experiment.

An auxotrophic mutant of *E. coli* that cannot synthesize arginine or thymine was grown in the presence of arginine and thymine and was then transferred to nutrient medium lacking arginine and incubated for several hours. Without arginine there is no net synthesis of proteins. In the absence of protein synthesis DNA replication cannot be initiated, and all the bacteria become arrested in the cell cycle at a point shortly before initiation of replication.

Addition of arginine to the medium allows protein synthesis to resume, following which all the bacteria begin to replicate their chromosomes at nearly the same time, that is, the millions of cells in the culture have been synchronised with respect to initiation of DNA replication.

At the moment of addition of arginine to the arginine-starved culture $^{3}$H-thymine of low specific activity was also added to the culture. (Low specific activity means that a small fraction of the thymine molecules possess a tritium atom.) When the cells begin replication, the incorporation of the low-specific-activity $^{3}$Hthymine results in a low level of labeling of the newly synthesized DNA chains, and a weak autoradiographic pattern will be produced by the DNA.

A minute or two after replication had begun, $^{3}$*H-thymidine* of high specific activity (roughly one $^{3}$H per thymidine) was added to the culture. Two things are important about the $^{3}$H thymidine. First, thymidine is a deoxynucleoside and is incorporated into DNA preferentially to thymine, the free pyrimidine base. Second, the addition of $^{3}$H-thymidine at high specific activity results in an immediate shift to incorporation of an intense level of radioactivity at replication forks and a resulting intense autoradiograph from the DNA. After a minute or two of incorporation of $^3$H-thymidine, the bacteria were cooled to stop further DNA replication, treated with lysozyme to digest the cell wall, and a drop of the culture was lysed with a drop of ionic detergent on a microscope slide. The detergent stripped the proteins from DNA.

The drop of culturedetergent mixture was spread, which caused DNA to be extended out onto the microscope slide. Autoradiography then revealed the pattern of radioactive regions in DNA molecules, as shown in Figure elsewhere in this chpater. Each molecule produced a central light autoradiograph for the two partial daughter segments ($^3$H-thymine incorporation).

The light pattern represented the first DNA to replicate after initiation. The intense autoradiographic patterns represented subsequent synthesis and showed that replication had proceeded bidirectionally from the origin (the origin is at the center of the light autoradio-graph). The two partial daughter molecules were joined at the replication forks at the outer extremities of the autoradiographic patterns. The unreplicated, parental DNA ahead of the replication fork was not radioactive and hence not seen in the autoradiograph.

Autoradiography shows the overall pattern of replication but does not have the resolving power to reveal the finer molecular events such as backfilling with Okazaki fragments, which occurs within the first micrometer behind the replication fork.

During replication, proteins that bind to DNA to produce the folding pattern of the chromosome, as well as proteins involved in transcription (RNA polymerase, repressor molecules, etc.) are probably transiently displaced from the DNA as the replication forks move through it. New proteins that make up the replication enzymatic machinery as well as single-stranded binding protein (SSB) bind to the DNA in the immediate region of the replication fork and migrate with it.

The SSB binds only to single-stranded DNA and stabilises the lagging parental strand prior to its replication. Behind the fork the newly replicated segments of daughter double helices become folded and packed by binding with packing proteins and may resume transcription with binding of RNA polymerase and other transcription factors.

All of these events-release of packing proteins, displacement of transcription factors, opening the double helix, and replication-occur extremely rapidly. At its maximum rate a replication fork moves through 980 by of DNA per second.

***Termination of replication***

The two replication forks, traveling in opposite directions, eventually meet (in 40 minutes at the maximum rate of DNA replication) on the opposite side of the circular chromosome. Figure elsewhere in this chapter shows an autoradiograph of two replication forks approaching each other less than a minute before termination. Termination apparently does not occur at a specific deoxynucleotide sequence.

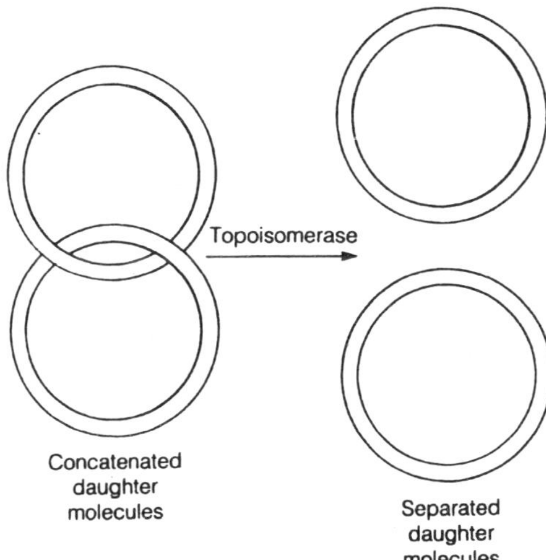


*Figure 8.10: At the end of replication of a circular duplex the two daughter duplexes are concatenated. The two molecules can be separated by topoisomerse, which can break and reseal DNA chains.*

Instead, in the termination region the rate of fork travel is enormously slowed (by an unknown means), and forks always meet in roughly the same region of the chromosome. The two replication forks abolish each other on meeting. At this point the two circular daughter molecules are complete, but are still interlooped with each other in concatenated fashion. The final step in replication is separation of the loops. Separation is catalysed by a topoisomerase, which creates a transient doublestranded break in one of the daughter double helices, allowing the other to slip through, and then reseals the double helix.

The connection between the termination of DNA replication and subsequent cell division is discussed in othere Chapter of this book.

## EUKARYOTIC CHROMOSOMES

The structure and replication of eukaryotic genomes are more complicated than in prokaryotes. Although there are important similarities between the two groups of organisms, there are also major differences.

Three are given here and others listed later.

1. In all prokaryotic organisms the entire necessary set of genes (genome) is carried in one chromosome. In contrast, almost all eukaryotic species distribute their genome over a number of chromosomes, ranging up to hundreds in some species.
2. The aggregate molecular weight of the DNA in a haploid set of chromosomes in a eukaryotic cell is a few times to tens of thousands of times greater than the molecular weight of the DNA in the largest prokaryotic chromosome known. The greater amount of DNA underlies the greater genetic complexity of eukaryotes, although much of the DNA in most eukaryotes has no known function.

**Table 8.2: Haploid Chromosome Numbers and Total DNA Content in Some Eukaryotes.**

| *Organism* | *Chromosome some number* | *Total Haploid DNA content in pg (C-value)* |
|---|---|---|
| Ant (*Myrmecia pilosula*) | 1 | — |
| *Haplopapus* | 2 | — |
| *Tetrahymena* | 5 | 0.23 |
| Budding Yeast | 17 | 0.016 |
| Nematode (roundworm) *Caenorhabditis elegans* | 6 | 0.09 |
| Chinese Hamster | 11 | 3.3 |
| Golden Hamster | 22 | 3.7 |
| Dog | 39 | 2.8 |
| Human | 23 | 3.1 |
| Salamander *(Amphiuma means)* | 12 | 100.0 |
| Lungfish *1Prolopterus)* | 17 | 50.0 |
| Lily *(Lilium)* | 12 | 50.0 |
| Turtle *(Chelonia)* | 28 | 2.7 |
| Boa Constrictor *(Constrictor)* | 18 | 1.8 |
| Frog *(Rana pipiens)* | 13 | 7.0 |
| Chicken *(Gallus domestica)* | 39 | 1.2 |
| *Alga (Euglena gracilis)* | 45 | 3.0 |
| Fruit Fly *(Drosophila melanogaster)* | 4 | 0.17 |

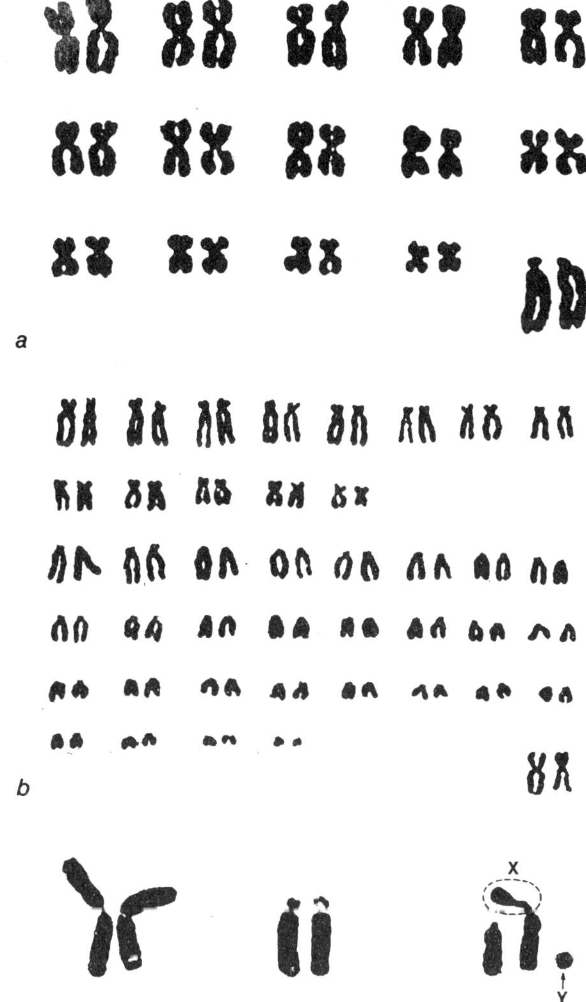


*Figure 8.11: Karyotypes of three mammals. (a) Chromosomes of a female Dorcas gazelle (Gazella dorcas); the diploid chromosome number is 30. The pair of X chromosomes is in the lower right corner. (b) Chromosomes of a female black rhinoceros* (Diceros bicornis); *the diploid number is 84, one of the highest numbers among mammals. The pair of X chromosomes is in the lower right corner. (c) Chromosomes of a male Indian muntjak deer* (Muntiacus muntjac). *The diploid chromosome number in males is seven and in females it is six, the lowest numbers among mammals. The X chromosome is normally attached to the third autosome, an unusual feature among mammals. The Y chromosome is very small. Although differing in chromosome number, all three of these mammals have about the same total amount of DNA in their genomes.*

3. The chromosome in a prokaryotic cell is always a covalently closed, circular DNA molecule. Each nuclear chromosome in a eukaryotic cell contains a single, linear molecule.

**Chromosome Numbers in Eukaryotes**

The smallest haploid chromosomal complement in a eukaryote occurs in a species of ant-the number is one. The plant *Haplopapus* has two chromosomes, and all of the thousands of other eukaryotes examined so far possess more than two chromosomes. Table elsewhere in this chapter lists the haploid chromosome numbers for a variety of eukaryotes, and Figure elsewhere in this chapter shows mitotic chromosomes of several organisms.

The chromosomal numbers show no regular pattern within groups of plants or animals. Different species of mammals have numbers that range from four in the Indian deer to 23 in the human to 39 in the dog. Mammalian species with fewer chromosomes have larger chromosomes. Some protozoa have as few as five chromosomes (e.g., *Tetrahymena);* others have hundreds *(Amoeba proteus)*. Large differences occur among plants, algae, amphibians, fishes, reptiles, and so forth, although in most species the chromosome number is less than 100. The significance of such divergent chromosome numbers, which occur even among closely related species, is not known.

**Amount of DNA in Eukaryotes**

The total amount of DNA in a haploid set of chromosomes in a species is called the genomic value or Cvalue of the species. The C-values for eukaryotes exceed the amount of DNA in the largest prokaryotic chromosomes by a few times (yeast) to several million times (some salamanders). This greater amount of DNA underlies the greater genetic complexity of eukaryotic species. C-values differ greatly among eukaryotes. For example, the C-value for the human is 3.1 pg, for the lungfish 50.0 pg., and for *Amphiuma means 100* pg. Even *Amoeba proteus* has many times more DNA than a human. It might be assumed that more DNA in the genome means a greater number of different genes. However, it can hardly be supposed that salamanders have more kinds of genes than humans. The resolution of this apparent contradiction between DNA amount and expected gene number, called the *C-value paradox*, is discussed later.

**The Size of DNA Molecules in Eukaryotes**

The amount of DNA per chromosome differs over a wide range in eukaryotes. The budding yeast has the smallest chromosomes discovered so far. The C-value of 1.5 × 10' base pairs of DNA is distributed among 17 chromosomes in a haploid set. The sizes of the DNA molecules in these chromosomes have been determined partly by electrophoretic migration in a gel. The smallest chromosome contains a DNA molecule of *260,000* base pairs (87 μm), or about 1/16 as long as the chromosome in *E. coli*. The others extend upwards in size with the largest one well over one million base pairs (over 333 Am in length).

**Table 8.3: Average Lengths of DNA Molecules in the Chromosomes of Some Eukaryotes.**

| ***Organism*** | ***Average Length of DNA Molecules*** |
|---|---|
| Yeast (*Saccharomvces cerevisiae*) | 0.3 mm |
| Unicellular Alga (*Chlamydomonas*) | 1.33 mm |
| Moss (*Sphagnum*) | 2.33 mm |
| *Fruit fly (Drosophila)* | 14.0 mrn |
| Chicken *(Gallus)* | 9.8 mm |
| Pike *(Esox)* | 3.1 cm |
| Human | 4.0 cm |
| Shark *(Scvlliur)* | 4.7 cm |
| Amoeba *(Amoeba proteus)* | 5.3 cm |
| Mouse (*Mus*) | 5.3 cm |
| Cat *(Felis)* | 6.2 cm |
| Pine tree *(Pinus)* | 9.7 cm |
| Frog *(Rana)* | 18.3 cm |
| Lungfish *(Protopterus)* | 98.3 cm |
| Onion *(Allium)* | 1.1 meters |
| Salamander *(Amphiuma)* | 2.7 meters |
| Lily *(Trillium)* | 4.3 metets |

There is no conceptual difficulty in accepting the length measurements for yeast chromosomes; they are all considerably smaller than the DNA molecule in *E. coli*. Other eukaryotes possess much more DNA per chromosome, in some the equivalent of DNA molecules several meters long. Such long molecules seem improbable, and for years it was debated whether a eukaryotic chromosome contains a single, long DNA molecule or multiple short ones.

The most direct way to determine whether the DNA in a chromosome is present in one or many molecules is to purify the DNA and measure its size directly by some physical technique. This has been done for the relatively short molecules in yeast, but is virtually impossible for much longer molecules because of the ease with which they are broken, especially during purification. Indirect methods for determining molecular length had to be devised as described below.

The lengths of the DNA molecules in chromosomes of various species cover a wide range. Table elsewhere in this chapter gives the average length of DNA molecules for some eukaryotic species. Why organisms package their genes in such vastly different sizes of DNA molecules is not known.

***Length of DNA molecules in Drosophila***

One method for measuring DNA length exploits an elastic property of DNA. Molecules of DNA free in solution coil randomly, much as a long piece of string coils when immersed in water. When a DNA molecule is pulled into a straight configuration and then released, it recoils back into a random configuration. The time taken for a straight molecule to relax is proportional to its length. Measuring the relaxation time provides an indirect but accurate measure of molecular length.

A great advantage of the method is that the DNA does not need to be purified to be measured, which eliminates manipulations of DNA that inevitably break it. Cells are lysed with an ionic detergent, which strips all the proteins from DNA. The elasticity of DNA is measured directly in the detergent solution.

When applied to cultured cells of *Drosophila,* the largest DNA molecule gives a molecular weight of $41 \times 10^9$, which corresponds to a length of 20.5 mm. The total amount of DNA in the largest chromosome was previously known from optical measurements to be $41 \times 10^9$, and the elasticity measurement proves that all this DNA is contained in a single molecule.

***Length of DNA molecules in mammalian cells***

Estimates of molecular length can be obtained for mammalian chromosomes with the autoradiographic technique previously described for visualising the DNA of *E. coli.* Cultured human cells are first grown for at least one cell cycle with $^3$H-thymidine, which completely labels the DNA molecules throughout their lengths. The cells are lysed in a drop of dilute ionic detergent on a microscope slide. Then the drop is gently spread over the slide, which draws the DNA into a long, straight form. Subsequent autoradiography shows the presence of tracks of silver grains several centimeters long, reflecting the presence of

DNA molecules of such length. A human diploid cell contains 6.2 pg of DNA and 46 chromosomes, which means on average an equivalent of 4 cm of DNA per chromosome. The autoradiographic visualisation shows that all the DNA in a chromosome is contained in a single molecule.

***Length of DNA molecules in salamanders***

The genomes and chromosomes of salamanders are among the largest known. Only the African lungfish, and a few species of lilies have comparable amounts of DNA per genome and per chromosome. A haploid set of 12 chromosomes in the salamander *Amphiuma means* (misnamed the Congo eel) contains 100 pg of DNA. Therefore, on the average each chromosome contains the equivalent of 2.7 meters of DNA. According to experimental evidence, some of which is cited below, all 2.7 meters of DNA are present in a single, extremely long molecule.

Most studies of DNA in salamanders have been done on chromosomes in meiosis. In salamanders the formation of oocytes lasts many months. During this time the chromosomes are arrested in the prophase stage of the first meiotic division. Because the chromosomes are partially condensed, they are easily seen by light microscopy.

A single lampbrush chromosome actually contains two homologous chromosomes that previously were tightly joined in meiotic pairing and subsequently have largely separated from each other. The chromosomes are now in the diplotene stage of the first meiotic prophase described in other chapter of this book.

The two homologues have separated except at several joined points called chiasmata *(sing.* = chiasma). Each chiasma is the cytological manifestation of a crossover event that took place earlier in prophase. Each homologue underwent duplication (DNA replication) in the premeiotic interphase and consists of two chromatids. Crossing over has occurred between a chromatid in one homologue and a chromatid in the other, which holds the chromosomes together at chiasmata.

The fuzzy appearance of the two homologues in a lampbrush chromosome is the result of threads that loop out from its two chromatids. The loops occur in pairs-wherever a loop occurs in one chromatid, it is matched by a loop of the same size in the other chromatid. The loops are made up of DNA with attached ribonucleoproteins and are sites of transcription. The axis of the chromosome consists of a string of fine chromatin granules called *chromomeres*. One or more pairs of loops extend from each chromomere. In this arrangement each chromatid consists of a string of chromomeres with attached loops. Since the two chromatids are tightly apposed to each other, the chromosome axis is probably formed by fused pairs of chromomeres.

The molecular structure of chromomeres and loops was deciphered by a combination of experiments. First, removal of proteins and RNA from lampbrush chromosomes with enzymes does not cause breakage of the chromosome. Treatment with DNase causes fragmentation. This suggests that the linear continuity of the chromosome is not dependent on protein or RNA but does depend on continuous intact DNA.

When protein and RNA are stripped from loops, only a DNA thread is left. Measurements of the diameter of the thread seen in loops

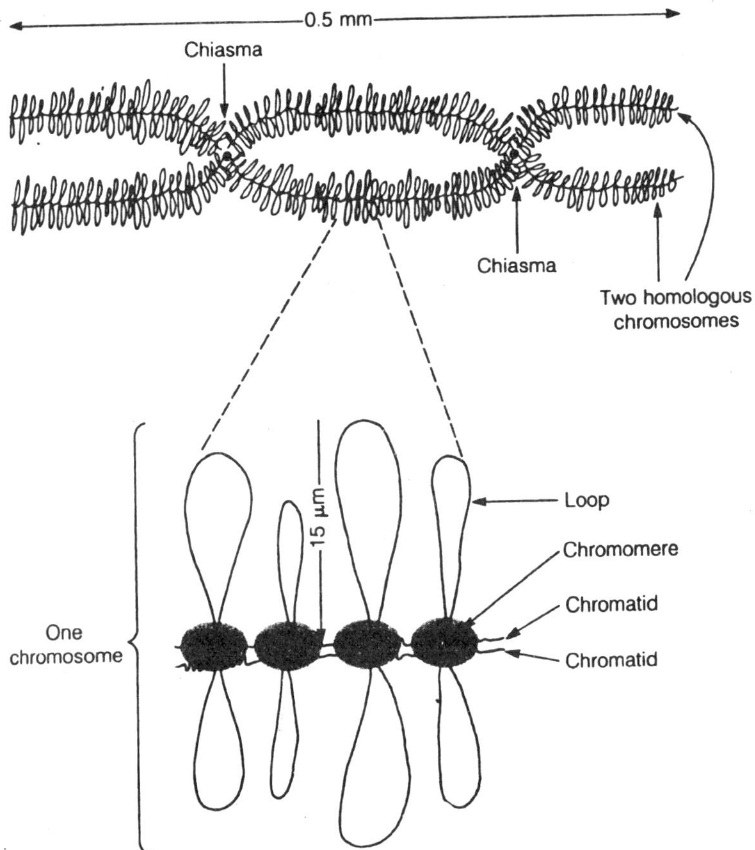


*Figure 8.12: Structure of a lampbrush chromosome.*

by electron microscopy show that the thread is a single DNA double helix. A comparison of the rates of breakage with DNase of loops versus the main axis of the chromosome shows that each loop contains one DNA duplex whereas the main axis must contain two DNA double helices, one for each chromatid.

These and other observations showed that each chromatid contains a single DNA double helix that extends through the entire length of the chromatid. Coiling and folding of the DNA gives rise to the successive chromomeres. Successive chromomeres are connected by segments of the DNA molecule that loop out from one chromomere and continue into the next. In short, one DNA molecule traverses the entire length of a chromatid. Proteins are responsible for folding the DNA in a specific pattern of successive chromomeres and loops.

The partial meiotic condensation of lampbrush chromosomes makes visible a structural organisation along the chromosome that may be present in interphase chromosomes of eukaryotes in general, although probably in a looser form. This presumed structural organisation is not discernible because the chromosomes are more extended and intertwined with one another within the nucleus.

Direct support for this view is provided by polytene chromosomes, described in Chapter 10. Polytene chromosomes are true interphase chromosomes. Their linear organisation is visible because many identical copies of the DNA double helix, with attached proteins, are held together in exact parallel register, producing a clearly discernible chromosome. Each band of a polytene chromosome may correspond to a chromomeric granule and its loop.

**Organisation of Deoxynucleotide Sequences in Eukaryotic DNA**

The DNA molecule of a prokaryotic chromosome is largely occupied by a succession of genes. A gene is defined here as a coding sequence and sequences upstream and downstream that are concerned with controlling the expression of the coding sequence. No genes have yet been found in some segments of the chromosome in *E. coli*, but most of the DNA consists of identifiable genes.

Eukaryotic DNA molecules are usually larger than prokaryotic molecules and possess a number of features not observed in prokaryotes. Earlier in the chapter three distinguishing features were mentioned to introduce eukaryotic chromosomes. Additional characteristics are the following:

1. The DNA of most eukaryotic genomes contains nongene

sequences that are repeated many times (*repetitive DNA*).

2. A few kinds of eukaryotic genes are present in multiple copies.
3. Eukaryotic DNA contains imperfect, apparently nonfunctional copies of some genes called *pseudogenes*.
4. The linear DNA molecules in eukaryotic chromosomes terminate in special sequences called *telomeres*, except for mitochondrial and chloroplast chromosomes, which are usually circular and therefore have no ends.
5. Eukaryotic chromosomes possess *centromeres*, structures by which chromosomes are distributed at mitosis and meiosis. The centromere has its basis in a specific sequence in the DNA molecule.
6. Prokaryotic DNA molecules contain a single origin of replication. The linear molecules in eukaryotes have many origins dispersed along their lengths.
7. Prokaryotic chromosomes are complexed with DNA-binding proteins that package the DNA molecule in successive units. Eukaryotic DNA is bound to a special class of proteins, the histones, forming repeating units called *nucleosomes*.
8. Most genes in eukaryotes are interrupted at a number of sites by introns. With a few exceptions introns are absent from prokaryotic genes.
9. Much of the DNA in most eukaryotic genomes forms long spacers between successive genes or groups of genes in a molecule. Spacers between genes in prokaryotes are very short.

***Repetitive DNA sequences***

All eukaryotes contain repetitive sequences. A small part of the repetitive sequences represents a few kinds of genes that are present in multiple copies, but most of the repetitive sequences have no known function, genetic or otherwise.

The absolute amount of repetitive sequences tends to be greater in organisms with larger genomes. In yeast cells only a few percent of the total DNA consists of repetitive sequences. Typically in mammals 30 percent to 40 percent of the DNA is made up of repetitive sequences belonging to several families. The remaining 60 percent to 70 percent is made up of sequences that occur only once per haploid genome. These are called *unique sequences* and contain most of the genes as well as a large amount of DNA of unknown function.

Within a family of repetitive sequences all the sequences are identical or nearly identical with one another and differ from sequences in any other family. Families are classified according to the number of copies of the repeat. Families containing two to 100 copies are *slightly repetitive*.

Families with 100 to several thousand copies are *middle repetitive*. Families with several thousand to several million repetitive sequences are called *highly repetitive*. Some of the families of slightly or middle repetitive DNA are, in fact, multiple copies of certain genes. For example, genes encoding actin are repeated several times in most eukaryotes and therefore make a family of slightly repetitive sequences. Genes encoding histone proteins (discussed later) are usually present in a few hundred copies and represent middle repetitive DNA. However, the functions, if any, of most families of middle repetitive sequences are unknown.

An example of a highly repetitive sequence occurs in the centromeric region of mouse chromosomes. This sequence is about 300 base pairs long, is repeated about $10^6$ times per genome, and accounts for 10 percent of the DNA in the mouse genome. It occurs in blocks of thousands of copics in tandem arrays. Its location in the centromeric region of chromosomes implies a centromere-related function, but its function is not known.

Blocks of highly repetitive sequences are packaged in a tightly condensed heterochromatin, which is in accord with the observed absence of transcription. Blocks of highly repetitive sequences often differ in overall base composition from the rest of the DNA in the genome. If the repetitive DNA has a higher-than-average content of AT base pairs compared with the rest of the DNA, it has a lower-than-average density in a CsC1 solution.

GCrich repetitive DNA has a higher average density. When total mouse DNA is cut into segments of a few hundred base pairs and centrifuged to equilibrium in a solution of CsCI, the highly repetitive DNA forms a small band adjacent to the main band of DNA because it has a higher average AT content and therefore a lower density than the main DNA. This, in fact, was how highly repetitive DNA was first discovered, and it is sometimes referred to as *satellite DNA* because it forms a satellite peak in a density profile.

The individual copies of sequences in families of all three classes of repetitive sequences (slightly, middle, and highly repetitive) can occur in blocks as just described for the family of highly repetitive sequences in the mouse genome, or the individual copies can be dispersed

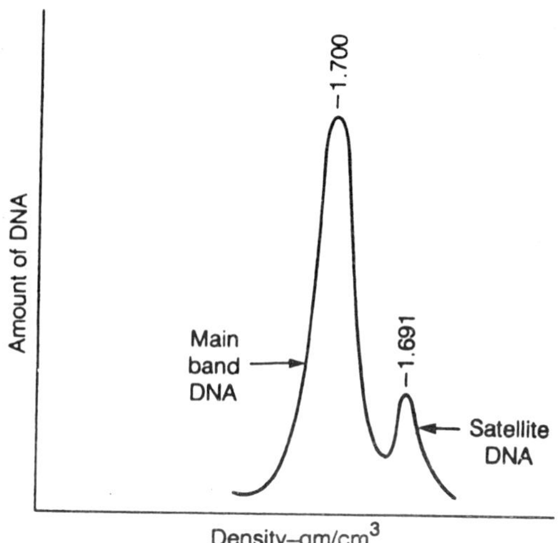


*Figure 8.13: Profile of mouse DNA centrifuged to equilibrium in a CsCI gradient. The DNA separates by density into two components, main-band DNA at a density of 1.700 gm/cm³ and a band of satellite DNA at a density of 1.691 gm/cm³.*

apparently randomly among genes throughout the genome. Middle repetitive DNA is also usually distributed in the genome in between genes.

An example of a highly repetitive sequence that is dispersed is the *Alu family* in the human genome. The repetitive sequence in the Alu family is about 300 base pairs long, more than 900,000 copies are present, and they account for five percent of human DNA.

One or more copies have been found not far from all genes isolated. At least some Alu sequences are transcribed, forming part of the primary transcript of some genes. The Alu segments of primary transcripts are removed and destroyed in processing of the transcript in the formation of an mRNA molecule. Like all repetitive sequences the function of the Alu sequences or their transcripts is not known. Closely related species sometimes have greatly differing amounts of repetitive DNA.

For example, the genome of *Drosophila melanogaster* contains about 15 percent of highly repetitive sequences, *D. virilis* has 40 percent, and *D. cyrtoloma* has 60 percent. In one species of the kangaroo rat 60 percent of the DNA is repetitive; in a closely related species less than ten percent is repetitive, and the total amount of DNA in the genome is correspondingly less.

***Telomeric DNA sequences***

The linear DNA molecules of eukaryotes have special sequences at their ends called *telomeric sequences*. The sequence of telomeric DNA has been determined in a number of lower eukaryotes. Although there are differences, all telomeric sequences conform to a common pattern. *In Tetrahymena* and *Paramecium,* telomeres consist of up to 70 repeats of the sequence 5'CCCCAA3'.

In a different group of ciliates (the hypotrichs) telomeres consist of several repeats of the sequence 5'CCCCAAAA3'. In a flagellated protozoan the repeated sequence in telomeres is 5'CCCTAA3', in yeast it is more variable, consisting of 5'CCACA3' and slight variations of the sequence. The telomeric sequences are important in conferring stability on linear DNA molecules, since experimentally introduced linear DNA molecules that lack them are degraded.

***Autonomously replicating sequences***

An *autonomously replicating sequence* or *ARS* is so named because it confers autonomous replicative capacity on a circular plasmid in yeast cells. It is comparable in function to the *on* sequence in a prokaryotic chromosome. The ARS in the yeast plasmid can be deleted, and a segment of foreign DNA inserted in its place to test for ARS activity of the foreign DNA in yeast cells.

Segments with ARS activity have been found in this way in chromosomal DNA of yeast and other organisms including human DNA, and these may be sequences defining origins of replication in their normal locations. The ARSs in yeast chromosomes differ somewhat from one another in sequence, but all those isolated so far have an 11-base-pair, AT-rich sequence with at least 200 base pairs flanking each side, which are necessary for normal initiation of replication. An ARS presumably is recognised by a specific protein that triggers the initiation of DNA replication, but such a protein has not yet been identified in eukaryotes.

***Centromeric sequences***

The proper distribution of chromosomes in eukaryotes during mitosis and meiosis is achieved by attachment of chromosomes to microtubules in the spindle apparatus in the dividing cell. The attachment in most species occurs at a region in the chromosome called the centromere. The centromere is also the point at which the two chromatids remain attached to each other in metaphase.

Separation of the two chromatids at the centromere marks the transition from metaphase to anaphase; from that instant on, the chromatids are called *daughter* chromosomes. The attachment of microtubules occurs at a part of the centromere called the *kinetochore*, which in the electron microscope looks like a plate on each chromatid. The centromere is defined by a sequence of deoxynucleotides. Special proteins probably bind to the centromeric sequence and give the centromere its functional properties and form the kinetochore.

Centromeric sequences from several chromosomes (*CEN sequences*) have been studied closely in yeast. They are quite similar but not identical in sequence, consisting of 80 to 90 base pairs of 93 to 94 percent AT base pairs next to a highly conserved 11-base-pair sequence. A CEN sequence can be removed from a chromosome and reinserted in the opposite orientation, and it works just as efficiently. CEN sequences, with slightly different composition, can be interchanged among yeast chromosomes without any loss of function.

***Pseudogenes***

*Pseudogenes* are nonfunctional, imperfect copies of normal genes. They occur coinmonly in higher eukaryotes such as mammals, but only a few have been found in lower eukaryotes such as yeast and *Drosophila*. In mammals pseudogenes account for as much as one percent of the total DNA. Formation of a pseudogene is a rare event in evolution, but once formed a pseudogene remains even though it is nonfunctional. For example, pseudogene Jn is a defective gene for β-globin of hemoglobin in mammals. A detailed evolutionary history of pseudogene $\psi\eta$ has been constructed by the study of the globin gene family in various animals, and it is clear that pseudogene $\psi\eta$ arose more than 140 million years ago, long before the evolution of modern mammals.

Pseudogenes arise by two mechanisms. Some, such as $\psi\eta$, originated by chance duplication of a normal functional gene. Duplication of a gene can occur by unequal crossing over during meiosis such that one chromatid receives both copies of a gene and the other chromatid receives no copy.

Gene duplication can also occur through a mistake in DNA replication. A segment of DNA is replicated twice, and the extra copy becomes integrated into a chromosome by a poorly understood mechanism of breakage and rejoining. The result is the presence of two functional copies of a gene.

Duplication of a gene may be followed sometime later in evolution by mutation of one of the copies into a nonfunctional form, creating a pseudogene. Defects observed in pseudogenes include abnormal initiation and termination codons, missense, nonsense, and frameshift mutations, altered promoters, and abnormal splicing in transcripts.

In the second mechanism pseudogenes are formed from mRNA

molecules. Certain RNA viruses, called *retroviruses*, carry a gene encoding an enzyme that can synthesize double-stranded DNA using a singlestranded RNA chain as a template instead of DNA. The enzyme is called *reverse transcriptase*.

The resulting segment of DNA double helix, called cDNA, may become integrated into chromosomal DNA by breakage and resealing. The integrated cDNA, being a copy of mRNA, lacks a promoter and other sequences necessary for transcription and hence cannot be transcribed. Pseudogenes formed from mRNA lack the introns present in most normal genes, and are therefore easily distinguished from their normal, intron-containing counterparts.

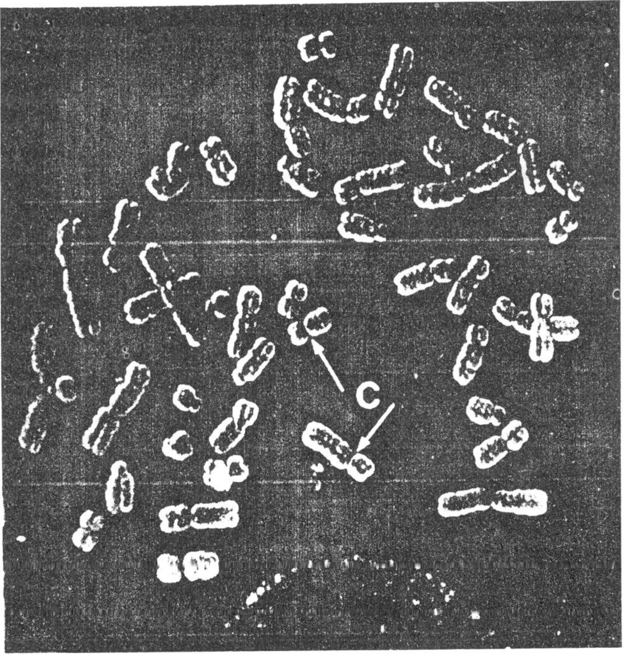


*Figure 8.14: Electron micrographs of kinetochores. (a) Metaphase chromosome with a platelike kinetochore on each chromatid. Microtubules from opposite poles of the cell are attached to the kinetochores. Most of the chromosome is out of the plane of the section. (b) Kinetochore on a daughter chromosome in artaphase. The chromosome is moving in the direction of the microtubules.*

Sequence analysis suggests that some families of dispersed repetitive DNA sequences are, in fact, pseudogenes of tRNA molecules, snRNA molecules, or other small non-mRNA molecules that are plentiful in the cell and therefore more likely to have been used by chance to make a

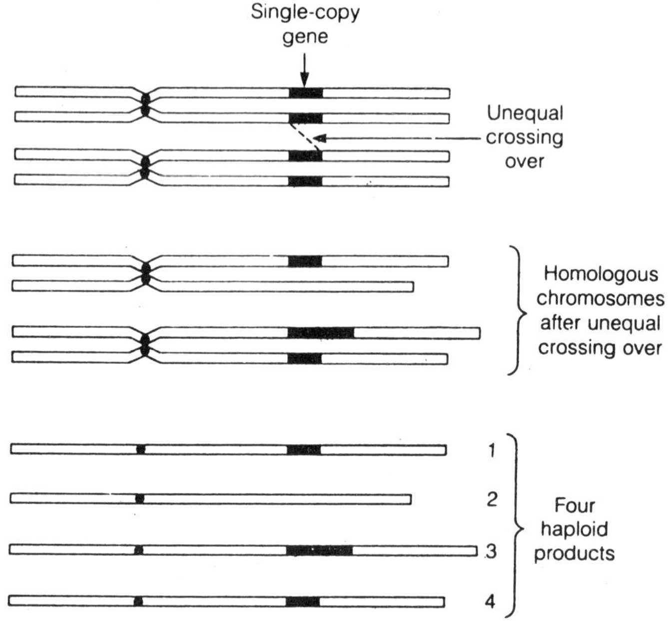


*Figure 8.15: Duplication of a gene in a chromosome by unequal crossing over. An organism receiving the third haploid product at fertilisation would contain a duplicate copy of the gene in one of its chromosomes.*

cDNA. Thousands of nonfunctional, pseudogene copies of snRNA genes are found in mammals. The human genome contains about 20 copies of the gene encoding actin; many of these are nonfunctional pseudogenes.

### *Junk DNA*

Pseudogenes, other dispersed repetitive sequences not clearly identified as pseudogenes, and perhaps other nonrepetitive sequences with no known function may all be defective relics of normal genes. For that reason they are sometimes called *junk DNA*. If such DNA sequences contribute nothing to the functioning of 'a cell, that is, are genetically neutral, it is not clear why they are not in the course of time lost through chance deletion. The cost of carrying junk DNA, for example, in the expense to the cell of its replication, may be so trivial that organisms containing it do not incur a survival disadvantage.

## The Essential Sequences of a Chromosome

Many kinds of sequences have just been listed; some are essential for chromosome structure and function, and some are not. The minimal

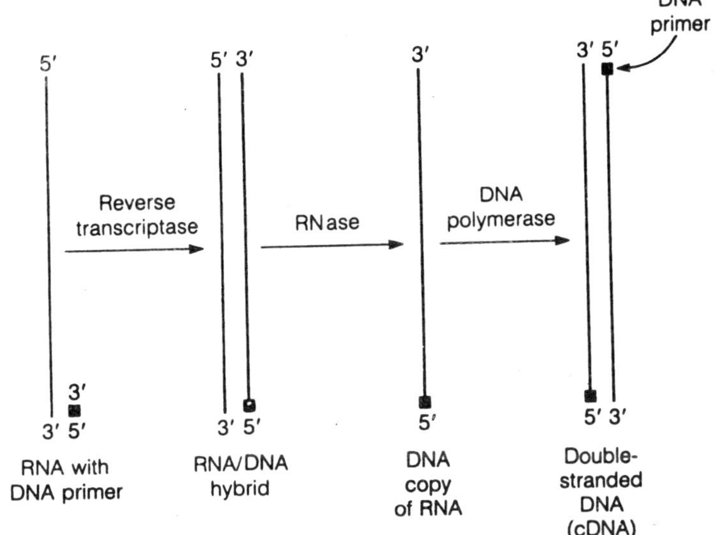


*Figure 8.16: Synthesis of a double-stranded DNA molecule starting with a single-stranded RNA template. Reverse transcriptase can use single-stranded RNA as a template, but like DNA polymerase it requires a DNA or RNA primer. In this case a short fragment of DNA is the primer. The RNA/DNA hybrid synthesized by reverse transcriptase is reduced to a single-stranded DNA molecule by RNase. DNA polymerase converts the single-stranded DNA into double-stranded DNA (cDNA), which requires another primer.*

sequences in a chromosome are genes, telomeres, ARSs, and a centromere. So far, research has hot identified functions for most repetitive sequences (not including normal gene families) and pseudogenes, and they are presumably nonessential.

*Figure 8.17: Essential sequences in a eukaryotic chromosome are those that are present in the two telomeres, the centromere, and an origin of replication. Multiple origins of replication are normally present, as well as genes and spacers between genes that give the chromosome a length required for reliable segregation at mitosis.*

The sequences required for chromosome function have been defined experimentally in yeast cells in the following way: a linear piece of DNA containing an ARS and a few genes was combined with a CEN sequence, and telomeres were joined to the ends. The molecule, containing about $10^4$ bp, was introduced into yeast cells. Although the molecule persisted through several cell divisions, it was lost because it was not properly distributed to daughter cells by mitosis. Apparently, the CEN sequence did not function adequately in such a short DNA molecule. A major difference between this artificially constructed chromosome and a normal yeast chromosome is length. The smallest yeast chromosome contains $2.6 \times 10^5$ bp, compared to the $10^4$ by in the artificial chromosome. The artificial chromosome was therefore enlarged to $5 \times 10^4$ by using DNA from a bacteriophage.

Bacteriophage DNA is functionless in yeast and was used to act as a filler. Increasing size in this way markedly improved the mitotic stability of the artificial chromosome and even allowed it to behave normally in meiosis, but it was still not as stable as the larger normal chromosomes in yeast. A normal yeast chromosome is accidentally lost with a frequency of once per $10^4$ to $10^5$ cell divisions. Smaller chromosomes are lost more frequently than larger ones.

These experiments imply that a DNA molecule, in addition to possessing ARS, telomeres, genes, and a CEN sequence must have some minimum size in order to be distributed properly in mitosis and meiosis. The apparent size requirement is met in yeast almost exclusively by genes, with little or no contribution from repetitive sequences, pseudogenes, or junk DNA.

**Total Complexity of DNA Sequences in Genomes**

The total complexity of sequences, as opposed to total DNA, is the sum of all the different sequences in a genome. In prokaryotes total sequence complexity is essentially equivalent to the DNA content of the genome since almost all the DNA consists of singlecopy, or unique sequences.

In eukaryotes, total sequence complexity is less than the total DNA content because of the presence of repetitive DNA. Repetitive DNA accounts for a significant part of the total DNA but because it is composed of the same sequence repeated over and over, it contributes disproportionally less than unique-sequence DNA to total sequence complexity. For example, about 25 percent of the DNA in the mouse genome is repetitive and 75 percent is unique. Of the 25 percent repetitive DNA, 10 percent is a highly repetitive sequence of 300 by (the satellite DNA), and 15 percent is middle repetitive DNA.

The total DNA content in the mouse genome is about $3 \times 10^9$ bp. The unique-sequence DNA (75 percent of total or $2.25 \times 10^9$ bp) accounts for virtually 100 percent of the sequence complexity. The highly repetitive sequence accounts for 10 percent of the total DNA but only 0.000013 percent of the total complexity (300 bp/$2.25 \times 10^9$ bp). The several families of middle repetitive DNA account altogether for less than one percent of the total sequence complexity in the mouse genome. The sequence complexity of DNA is determined by shearing it into fragments and dissociating the two chains of the double helix by denaturation; then, the rate at which they reassociate is measured as a function of DNA concentration expressed as nucleotides per liter. The greater the complexity of DNA, the more slowly it reassociates. Consider, for example, two DNA molecules each consisting of 1000 bp.

Suppose one of the molecules consists entirely of unique sequences and the other consists of a 100-base-pair sequence repeated identically ten times. Suppose the same number of molecules of 1000 by is dissolved in a salt solution of the same volume, cut in segments about 100 by long, denatured, and then allowed to renature. The rate of reassociation is governed by how often complementary chains encounter each other in solution. The concentration of chains of particular sequences is ten times higher for the repetitive DNA than for the unique sequence DNA; therefore the repetitive DNA will reassociate much faster than the unique sequence DNA. Likewise, when the total DNA of a genome is cut into lengths of a few hundred by and dissociated, each kind of sequence reassociates at a rate that is proportional to its concentration; highly repetitive sequences reassociate more rapidly, followed by reassociation of middle repetitive sequences, and then by reassociation of the unique sequences.

Figure elsewhere in this chapter shows reassociation curves for DNAs of T4 bacteriophage, *E. coli,* and the mouse. Since T4 and *E. coli* DNA do not contain repetitive sequences, they reassociate as uniform curves reflecting rates expected for genomes of their respective sizes. The genome of *E. coli* DNA is about *25* times larger than DNA of T4. If both DNAs are prepared at the same concentration (in nucleotides per liter), the DNA of *E. coli* reassociates more slowly than T4 DNA (because the number of DNA molecules is less).

Mouse DNA shows a more complicated curve, the first part of which reflects reassociation of highly repetitive sequences (satellite DNA), followed by several families of intermediate repetitive DNA and finally unique sequence DNA.

A useful way to record reassociation events is to plot the initial molar concentration of nucleotides ($C_o$) in the DNA times the time (t) in seconds that reassociation has been occurring against UV absorption; a decline in UV absorption occurs as DNA reassociates. Such a plot of initial $C_o \times t$ (Cot) against UV absorption is shown in Figure elsewhere in this chapter. The UV absorption of totally dissociated

DNA at t = 0 is defined as 100 percent. The value of Cot when half the DNA has reassociated, $Cot_{1/2}$, is used as a measure of the sequence complexity of the DNA in nucleotides. Notice in Figure *11-25* that the $Cot_{1/2}$ for *E. coli* DNA is about *25* times greater than for T4 DNA, and the $Cot_{12}$ for the *unique fraction* of mouse DNA is about 600 times greater than for *E. coli* DNA.

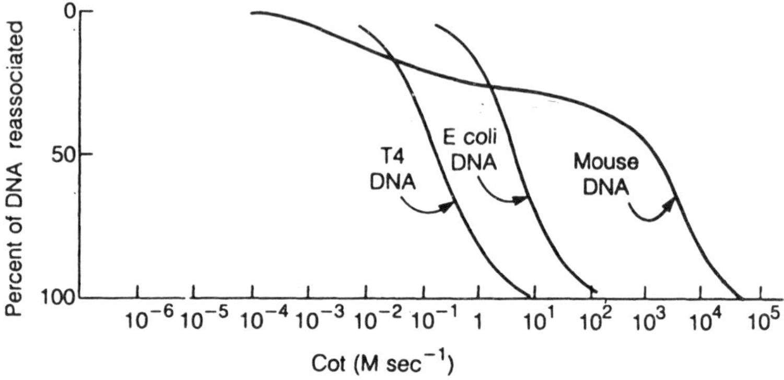


*Figure 8.18: Reassociation (Cot) curves for bacteriophage T4 DNA, E. coli DNA, and mouse DNA.*

The measure of sequence complexity of genomes, or *S-value* (S for sequence), is in some ways a more useful value than is total DNA since total DNA can differ by wide margins from species to species because of large differences in the amounts of repetitive sequences. The DNA content of mammals covers a twofold range, but all have about the same complexity, which is about 600 times the complexity of *E. coli* DNA. This does not mean that mammals possess 600 times more genes than *E. coli* because most of the DNA sequences in mammals do not encode genes.

### The C-Value Paradox

The C-value paradox mentioned earlier occurred because of wide differences in DNA content among species. Comparing S-values among organisms instead of comparing C-values might be expected to show a closer correspondence. After all, S-values are measures of the amount of unique sequence complexity, and they avoid the wide differences in total DNA that stem from differences in the content of repetitive DNA. The correspondence is, indeed, closer when S-values are used. All mammals have close to the same structural and functional complexity, and although C-values vary over a two-fold range, they all have very close to the same S-value. However, among eukaryotes a lack of correspondence between S-values is still often observed. For example, salamanders have several times more unique-sequence DNA than do mammals. The solution to this S-value paradox is that a large amount of unique-sequence DNA does not encode genes, as described in the next section.

**Estimates of Gene Numbers, the S-Value Paradox, and Spacer DNA**

The precise number of genes is known for a few of the simpler viruses, but it is not known for any prokaryotic or eukaryotic cell. Assuming 1000 by for one gene, mycoplasma contains a maximum of 750 genes. The chromosome of *E. coli* contains $4.7 \times 10^6$ bp, enough DNA for 4700 genes, although the number is certainly much less than that. Eukaryotes contain more genes than prokaryotes, but at present only rough estimates are possible.

***Gene number and spacers in Drosophila***

*Drosophila melanogaster* contains $1.65 \times 10^8$ by of DNA, of which 85 percent or $1.40 \times 10^8$ by is unique sequence DNA. Assuming a somewhat larger average size for the coding regions of genes in *Drosophila,* such as 1200 bp, $1.40 \times 10^8$ by could contain 117,000 genes. Estimates based on other grounds are considerably less.

Genetic evidence suggests that one band of a polytene chromosome corresponds to one gene, although some bands may contain more than one gene. There are a little over 5000 bands total in the four polytene chromosomes, so estimates of gene number are 5000 to 7000. At 1200 by per gene there is a 17- to 23-fold excess of unique sequence DNA.

A clue about the disposition of the excess uniquesequence DNA is apparent from the study of the polytene chromosomes in *Drosophila.* On the average, bands in a polytene chromosome contain segments of DNA consisting of about 25,000 by of unique sequence.

Using the estimate of 1200 by for a gene, with one gene per band, about five percent of the DNA in a band is used to encode a gene. Some part of the remaining 95 percent (23,800 bp) can be accounted for by introns and pseudogenes. However, the function of the bulk of the unique-sequence DNA cannot be accounted for, and at present it must simply be regarded as spacer DNA between successive genes.

***Gene number and spacers in mammals***

In mammals the amount of unique sequence is sufficient to encode over $1.5 \times 10^6$ genes. A number of genetic considerations suggest a number 15 to 30 times less, or between 50,000 and 100,000 genes. Most eukaryotic genes contain introns, sometimes many of them, and some are very large. Even including introns as parts of genes does not resolve the S-value paradox. The solution is, again, probably the presence of large spacers of unique sequences between successive genes. For example, a 50,000-bp segment of DNA in humans contains the five members in the family of the, β-globin genes and the $\psi\eta$ pseudogene for β-globin.

Each gene consists of about 2/3 intron sequences and 1/3 coding sequences. However, the six genes, including their introns and known regulatory sequences upstream and downstream from the genes, account for only about 20 percent of the 50,000-bp segment. The rest of the DNA, most of which consists of unique sequences, contains no identifiable genes and appears to exist as spacers between members of the globin gene family.

### *Gene number and spacers in ciliated protozoa*

In ciliated protozoa it is clear that most of the uniquesequence DNA in the eukaryotic chromosomes exists as spacers between genes and has no known function. Ciliates were introduced briefly in other Chapter. A ciliate contains a small, diploid micronucleus and a large, DNA-rich macronucleus. The micronucleus does not produce RNA, and it functions during cell mating as described in other chapter of this book. The macronucleus makes all the RNA needed for cell maintenance and reproduction.

The major activity of ciliates is growth and division but occasionally two cells mate (conjugate). During conjugation the micronucleus undergoes meiosis, and the cells exchange haploid micronuclei and then separate, as described in othere chapter. Immediately after separation the macronucleus is destroyed, and a new macronucleus is formed from a micronucleus. Formation of a new macronucleus is a complex process that need not be described here.

In one group of ciliates known as hypotrichs, the first stage in macronuclear development is the formation of polytene chromosomes by multiple rounds of DNA replication. The process takes two days. As soon as they have formed, the polytene chromosomes are cut into short segments, each segment corresponding to a band of a polytene chromosome.

A band is believed to represent a single gene, and the cutting up of the polytene chromosomes is; therefore, a gene-by-gene process. Cutting produces segments of DNA that contain on the average over 40,000 bp. These are rapidly reduced to an average of 2200 by by cleavage of most of each DNA segment to nucleotides. Each of the short molecules that is left represents a single gene. Short telomeric sequences are then added to the ends of each gene. Finally, all the genes replicate many times to produce a mature DNA-rich macronucleus in which all the DNA occurs in short molecules, each encoding a single gene.

The mature macronucleus contains about 20,000 different DNA molecules, each present in about 1000 copies. These genes, representing only five percent of the sequence complexity in the chromosomes of the original micronucleus, are transcribed to provide all the mRNA, tRNA, rRNA, and other RNAs needed by the cell.

In summary, the development of the macronucleus consists of excision of all the genes from the chromosomes and destruction of most of the sequences in the chromosomes. Why polytene chromosomes are formed as an intermediate stage and many of the molecular details of how genes are excised and supplied with telomeres are still not known.

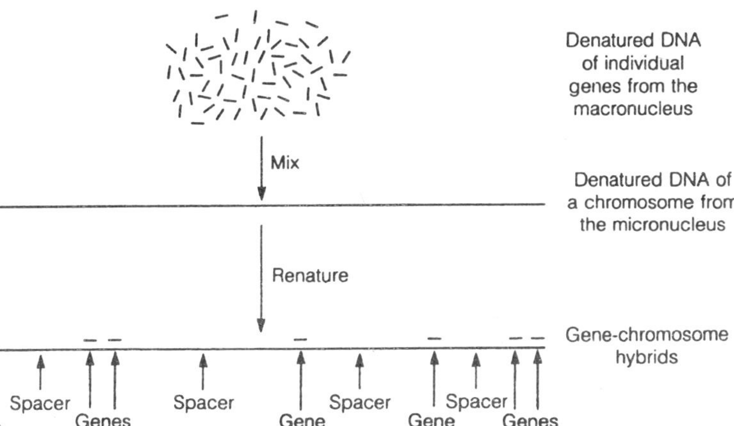


*Figure 8.19: Organisation of genes in a chromosome. Small DNA molecules containing single genes from the macronucleus are denatured and mixed with denatured DNA of chromosomes from the micronucleus. After renaturation, gene-sized molecules are found hybridised to chromosomal DNA in a highly dispersed pattern, with large spacers between genes or small groups of genes.*

However, the unusual manipulations of the genome by these ciliates reveal how gene and nongene sequences are organised in chromosomes. The gene-sized molecules in the macronucleus can be used as hybridisation probes to map the position of the genes in chromosomal DNA using long segments of DNA cloned from micronuclear chromosomes.

Genes are found singly or in small groups along the DNA molecule separated by long spacers of nongene DNA. It is the spacer DNA, accounting for 95 percent of the total deoxynucleotide sequences, that is degraded following the gene-by-gene segmentation of the polytene chromosomes. This extreme elimination of all nonessential sequences to form the functional nucleus presumably provides some advantage to these organisms.

Ciliates may require the many copies of each gene in the macronucleus to support their very large cell size. By elimination of all nonessential sequences, a high copy number for each gene can be achieved with a minimum amount of DNA and a minimum expansion of nuclear size.

The several observations on the DNA of *Drosophila,* mammals, and ciliates are consistent in suggesting a general model for eukaryotes, in which most of the DNA sequences form spacers of unknown significance between successive genes. Large differences in DNA sequence complexities (S-values) among eukaryotes, such as the high values in salamanders and lilies, are probably due to large differences in the size of spacers between genes.

**Gene Families**

Most genes in both prokaryotes and eukaryotes are present in single copies per genome. Although a few kinds of genes exist in multiple copies, in general single copies of genes are sufficient, because expression of genes that encode protein products is amplified at two points. First, many mRNA copies are transcribed from a single gene, and second, each of the many mRNA copies is translated into many copies of the protein.

For most genes these amplifications at the transcriptional and translational levels are sufficient to meet the needs of a cell. Even during differentiation of cells, when a large amount of a gene product may be required, single copies of genes are sufficient. For example, an enormous number of hemoglobin molecules is produced in the formation of a red blood cell. Single copies of genes encoding α- and β-globin are sufficient.

For genes whose final products are RNA molecules, like genes encoding tRNAs and rRNAs, expression is by transcription alone. For these genes, as well as for a few kinds of genes that encode proteins, the rate of production of product cannot be achieved with a single copy of the gene per genome because of limits on rates of transcription (10 to 50 mRNA transcripts per gene per minute and 3 to 20 polypeptides per mRNA molecule per minute).

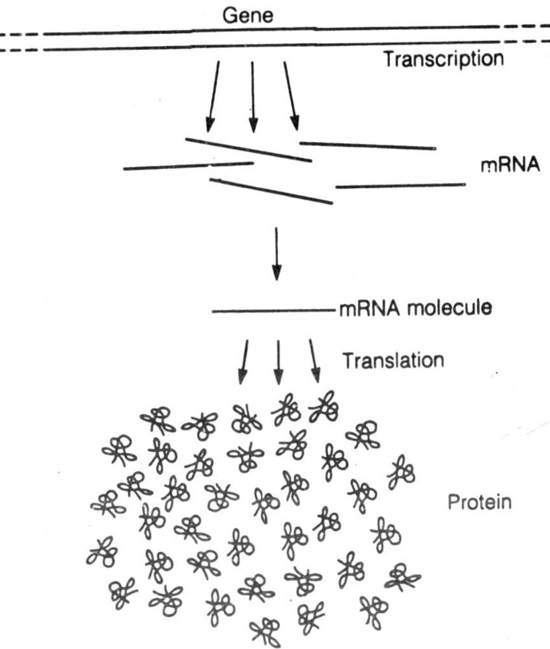


*Figure 8.20: Expression of protein-encoding genes is amplified by multiple transcriptions, with accumulation of mRNA molecules, and by multiple translations of each mRNA molecule, with accumulation of protein molecules.*

Almost all eukaryotes have multiple copies of the sequences encoding the four rRNAs. Sequences for three of the rRNAs (28S, 18S, and 5.8S in mammals) are grouped in a single transcription unit of DNA and are transcribed as a single precursor rRNA that is posttranscriptionally processed into three final products as described in other chapter.

Transcription units for the three rRNAs are repeated in tandem in the chromosome in many copies. In yeast cells the transcription unit is repeated 100 to 120 times in the arrangement shown in Figure elsewhere in this chapter. Successive transcription units are separated by about 2000 bp. Within the 2000 by is a gene of about 120 by that codes for the fourth rRNA, 5S rRNA.

The rest of the 2000 by is spacer. In other eukaryotes the 5S genes occur elsewhere in the genome as separate tandem arrays. In these eukaryotes the spacers between the transcription units for the three rRNAs contain no coding function, and no part of them is transcribed.

The size of spacers between rRNA transcription units ranges from about 2000 by in some protists to 50,000 by in some higher eukaryotes.

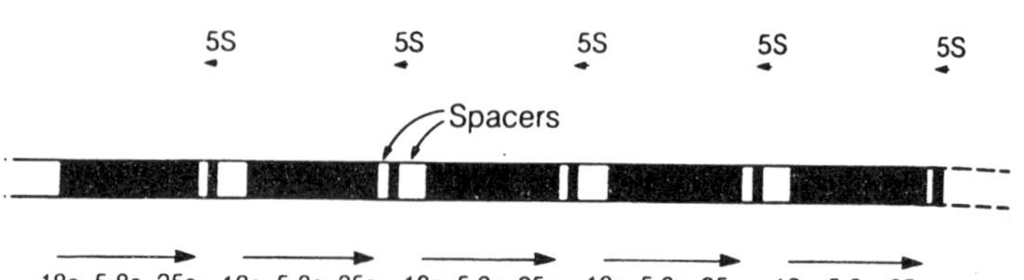


*Figure 8.21: Tandem arrangement of genes encoding 5S, 18S, 5.8S, and 25S rRNA in yeast. The genes encoding 5S rRNA are transcribed separately and in the opposite direction from the genes encoding 18S, 5.8S, and 25S rRNA sequences. Arrows indicate the direction of transcription.*

There are 450 rRNA transcription units per haploid genome in *Xenopus,* arranged in tandem. Each is 7600 by long and separated from adjacent units by untranscribed spacers of several thousands of base pairs. The transcription of rRNA genes and their separation by spacers can be observed directly by electron microscopy in gently lysed and dispersed nuclei.

The hundreds of genes for rRNA form the *nucleolar organiser,* the core of the nucleolus. In the nucleolus rRNA transcripts are processed into three ribosomal RNAs and combined with the fourth rRNA (5S) and with ribosomal proteins, which are synthesized in the cytoplasm and imported into the nucleus, to form the large and small ribosomal subunits. These are then exported to the cytoplasm.

Thus, the nucleolus is the structural manifestation of rRNA transcription, rRNA processing, and assembly of ribosomal subunits. Genes encoding 5S rRNA are arranged in multiple tandem clusters, usually in several different chromosomes and do not form part of the nucleolus. In *Xenopus* there are 24,000 copies of 5S genes that are separated by untranscribed spacers.

The genes are 120 by long and are separated by spacers of 600 bp. Genes encoding tRNA molecules are typically repeated about 200 times and are separated by spacers that are, on the average, 10 times longer than a tRNA gene itself. In growing oocytes of amphibians and some insects even the many copies of genes encoding 18S, 28S, and 5.8S rRNAs are insufficient to meet the need for ribosome production.

In these cells the genes for these rRNAs are extensively amplified further. Histones are proteins needed in the packing of long DNA molecules in the nucleus. There are five types of histones and 20 to several hundred copies of each gene are present. In most eukaryotes histone genes are arranged in groups, with one copy of each of the five genes per group. In sea urchin cells there are about 1000 such groups repeated in tandem. The five genes within a group are transcribed individually and separated by spacers. The groups in turn are separated from one another by spacers.

The spacers that occur between copies of repeated genes are usually similar or identical in sequence for a particular gene in a particular species. Thus, all the spacers between 5S rRNA genes are the same in *Xenopus laevis*. In the closely related species, *X. mulleri*, the 5S genes are identical to those in *X. laevis* but the spacers are three times longer and have a different sequence from the spacers in *X. laevis*. In general, spacers change in length and sequence far more rapidly during evolution than the genes they separate. Whatever the function of spacers, that function is compatible with large changes in spacer composition.

The significance of spacers is unknown. However, it is presumed that spacers between repeated genes represent the same phenomenon and have the same significance as spacers between single copy genes.

## Packing of DNA in Chromosomes

The length of the DNA molecules in eukaryotic chromosomes exceeds by many thousands the diameter of the nuclei in which they are contained. Extreme examples are the DNA molecules of some salamanders. Twelve DNA molecules, each several meters long, are contained in a spherical nucleus less than 20 Am in diameter in *Amphiuma*.

In humans a total of 200 cm of DNA in the 46 chromosomes of a diploid cell are contained in a nucleus with a diameter of 6 or 7 $\mu$m. By coiling and packing, the extremely long DNA molecules can fit into the small nuclear volume. Certain nuclear proteins, described below and in Chapter 10, bring about the coiling and packing process.

### *Nucleosomes and chromatin*

The DNA of eukaryotes is complexed with proteins that, in effect, package the DNA in stretches and hold it in specific configurations within the chromatin. The arrangement of proteins and DNA in chromatin is partially understood. Also, it is clear that the particular configuration of DNA in chromatin is very important in the regulation of transcription. In general DNA packaged tightly with protein in chromatin is not transcribed; loosening of the chromatin is necessary, but alone is insufficient to bring about transcription. For transcription to occur other proteins must interact with the DNA sequence upstream from the coding region of a protein-encoding gene.

The packing of DNA in chromatin occurs in several successive orders, each of which accomplishes a further compaction of the DNA. The primary level of packing has been thoroughly elucidated by biochemical and physical analyses and electron microscope observations on chromatin isolated from interphase nuclei. It has been known for a long time that DNA is complexed with a family of five histone proteins. Histones contain large numbers of the basic amino acids lysine and arginine, with R groups that terminate in amino groups ($—NH_2$). In the pH range in the cell the amino groups acquire a proton (H+) and are therefore charged positively ($—NH_3^+$).

The large number of positive charges of arginine and lysine residues in histones form ionic bonds with the negative charges of the phosphates in the DNA strands. In the last decade the molecular details of how histones are complexed with DNA and form the first level of chromatin organisation have been worked out.

Brief treatment with deoxyribonuclease (DNase) breaks chromatin into small particles, each composed of histones complexed with a fragment of DNA. After the histones are removed, the DNA consists of fragrhents of about 200 by and multiples of 200 by (i.e., 400, 600, etc.). If the enzymatic digestion is longer, all of the DNA is in units of 200 bp, and the multiple-size fragments are not found. These and other experiments show that the long, continuous DNA molecule in a chromosome is complexed with histones in 200-bp stretches.

Each stretch is complexed with histones so as to form a particle. In corollary observations by electron microscopy, chromatin that has loosened and spread out is found to consist of long strings of particles. Brief treatment with DNase breaks the string into units containing one to several particles, and longer digestion reduces the string entirely to individual particles. Each particle can be dissociated into a 200-bp fragment of DNA and a particle consisting of eight histone molecules.

The particles of chromatin, or *nucleosomes*, are joined in long strings because the DNA molecule is continuous from one nucleosome to the next. The region of DNA linking nucleosomes is more accessible to DNase than DNA within nucleosomes. Hence, with properly gauged digestion chromatin is converted into single, physically separate nucleosomes.

Each of these nucleosomes is made up of a 200-bp fragment of DNA and nine histone molecules-one molecule of histone H1, and two molecules each of the other four histones H2A, H2B, H3, and H4. If the nucleosome preparation is digested further with DNase, the DNA fragment is gradually reduced in size until it reaches 146 by and

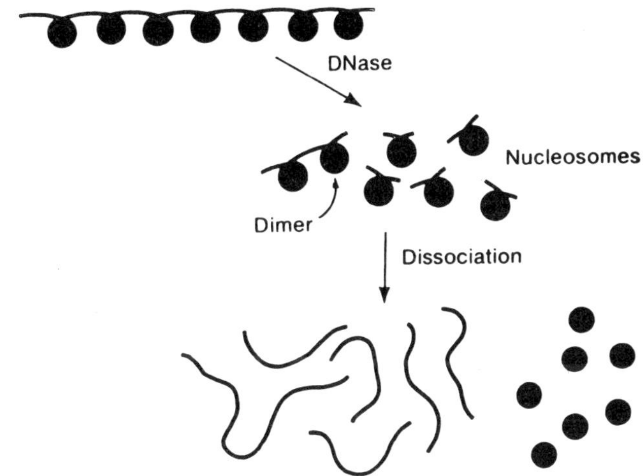


*Figure 8.22: Strings of nucleosomes digested with DNase form single nucleosomes, dimers, trimers, etc.*

histone H 1 is released. Further action of DNase is blocked because of a tight association between the remaining eight histone molecules and DNA.

The structure that remains is called the *core partide*. The eight histone molecules (two each of H2A, H2B, H3, and H4) in a core particle are joined with each other in a specific pattern that forms a disc. The DNA molecule makes almost two turns around the disc and then continues to the next core particle. The DNA extending from one core particle to the next is called *linker DNA*. While the amount of DNA wrapped around the histone aggregate is constant (.146 bp), the amount of DNA in the linker is variable from one kind of cell to another or from species to species. On the average, linkers are a little over 50-bp long, which, added to the DNA around the core particle (146 bp), accounts for the 200-bp fragments observed after brief DNase digestion of chromatin.

An H1 histone molecule is attached to each stretch of linker DNA; hence it is released when the linker DNA is digested with DNase. H1 histone plays a role in holding adjacent nucleosomes in tight association and in folding or twisting of the string of nucleosomes into the next order of DNA packing.

The packing of DNA into nucleosomes condenses the DNA about seven-fold, forming on the average for a human DNA molecule (four cm long) a thread of 600,000 nucleosomes that is about ten nm (1 nm = $10^{-9}$ meter) in diameter and 6000 Am long. Relative to the diameter of

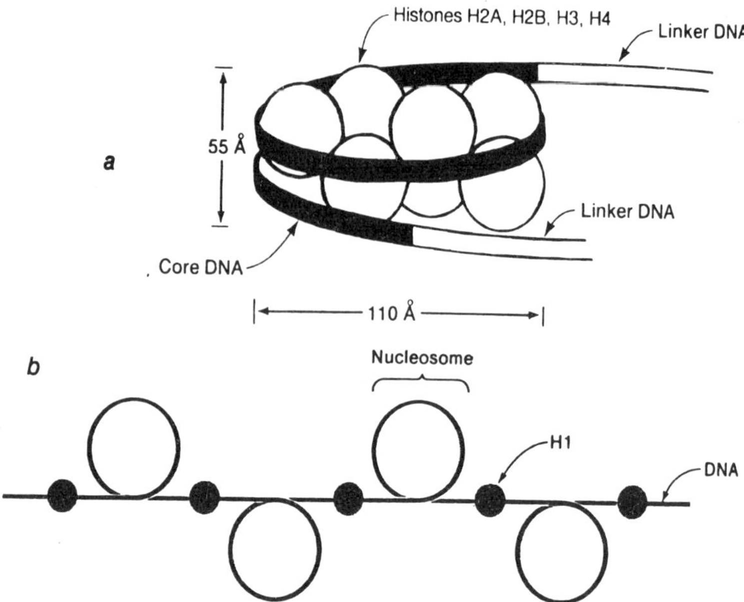


*Figure 8.23: (a) Diagram of a nucleosome core particle. The DNA molecule makes 1 3/4 turns (consisting of 146 base pairs) around a histone octomer (two molecules each of histones H2A, H2B, H3, and H4). The DNA continues through linker DNA into adjacent nucleosomes. (b) A string of nucleosomes, with histone H1 bound to the linker DNA between individual nucleosomes.*

the nucleus (6 to 7 $\mu$m), the string of nucleosomes is still enormously long.

In the next order of DNA packing, the ten-nm nucleosome thread is folded into a fiber about 30 nm in diameter. According to the most widely accepted model the nucleosome thread is arranged as a hollow helix or solenoid, generating a thicker structure.

The solenoid is believed to account for the thread of *30* nm in diameter seen by electron microscopy of chromosomes. The thicker filament formed in the solenoidal model is about seven times shorter than the extended nucleosome filament or a little over *800* t4m for an average human chromosome. Observations by electron microscopy suggest that at the next higher level of packaging the solenoidal filaments are arranged as loops, containing 5 to 10 $\times$ $10^4$ by of DNA in an animal cell, extending out from a central axis composed of protein and resembling the arrangement seen in lampbrush chromosomes.

The three orders of packing of DNA (nucleosomes, solenoids, and central axis with loops) accomplished by histones and other proteins is

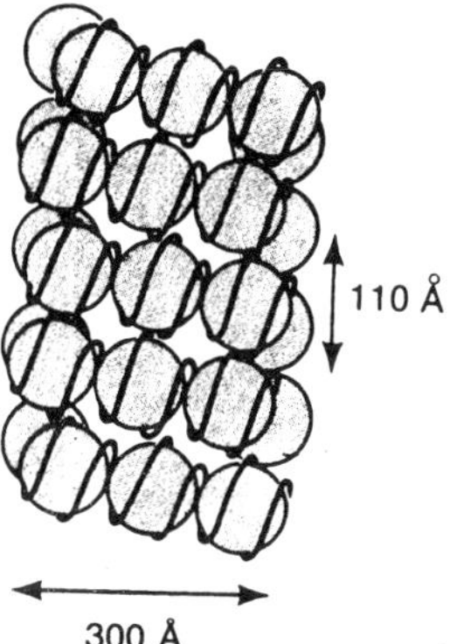


*Figure 8.24 : A solenoidal model of the packing of nucleosomes, yielding a fiber about 300 A (30 nm) in diameter. The DNA double helix (red) is wound around each nucleosome.*

thought to produce the chromatin observed in the interphase nucleus. This chromatin contains the DNA that is transcribed and forms euchromatin. A portion of the chromatin undergoes additional packing to yield the more tightly condensed form, heterochromatin. The molecular arrangements in this mode of packing are not known. The significance of euchromatin vs heterochromatin is not completely understood, but DNA present in heterochromatin is generally not transcribed.

### *Organisation of chromatin*

The contents of the nucleus appear to form a gel with no movement of the euchromatin, heterochromatin, or nucleolus. There is, for example, no Brownian movement (vibratory movements caused by molecular collisions) of intranuclear structures in living cells viewed by light microscopy. By comparison, the cytoplasm is largely a solution in which organelles like mitochondria and lysosomes visible in the light microscope are in constant Brownian movement. The immobilised state of interphase chromosomes also contrasts sharply with the constant Brownian movement activity and shuffling movements of chromosomes in mitosis.

Those parts of interphase chromosomes that are condensed into heterochromatin are often closely associated with the inside of the nuclear envelope. A striking example is the inactivated X chromosome in a female mammalian cell. The entire chromosome is packed as a discrete mass of heterochromatin, the Barr body, that is also positioned at the nuclear envelope.

The inside of the nuclear envelope is lined with a thin layer of protein, the ***nuclear lamina***. The lamina may serve to hold and therefore

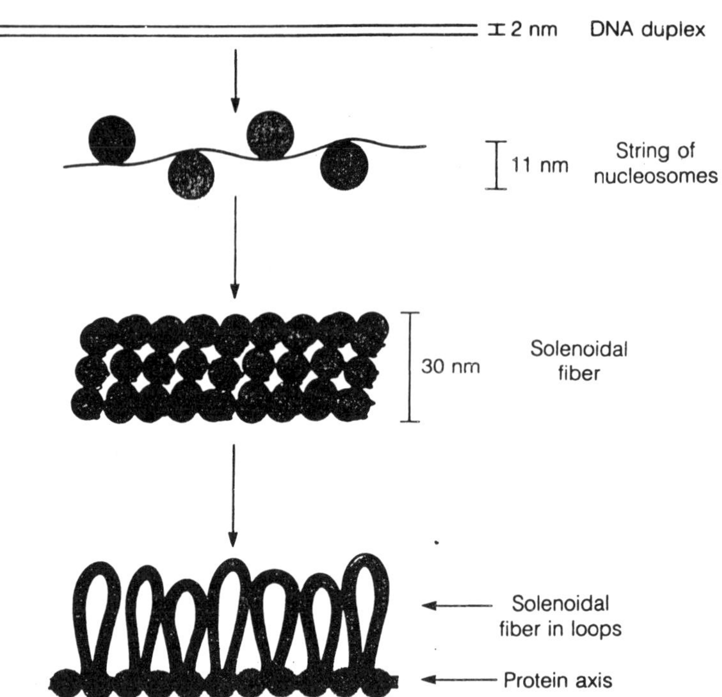


*Figure 8.25: Model showing three orders of packing of DNA in chromatin: (1) The DNA duplex is wound around particles of histone to form a string of nucleosomes 11 nm in diameter. (2) Nucleosomes are packed into a solenoidal fiber 30 nm in diameter. (3) A solenoidal fiber is arranged in loops, with loops anchored into an axis composed of proteins.*

probably mediate the adhesion of chromatin to the nuclear envelope. The two membranes of the nuclear envelope are continuous with one another as they form the walls of the many nuclear pores that are present in the nuclear envelope.

The inner and outer openings of a pore and the interior of the pore contain eight granules, arranged radially and surrounding a central granule that is sometimes present in the opening. The granules, together with the pore, are called the *nuclear pore complex*. The entire complex is about 100 nm in diameter. The particles of the pore complex together with some ill-defined fibrous material appear to occlude most of the pore opening. Nuclear pores are nevertheless believed to be channels by which materials move between nucleus and cytoplasm. When molecules with molecular weights of less than 50,000 are injected into the cytoplasm, they quickly equilibrate between the nucleus and cytoplasm. Larger molecules enter the nucleus slowly. From such experiments it has been estimated that the effective diameter of the opening of the pore is about nine nm.

Molecular traffic through nuclear pores is intense. For example, during DNA replication hundreds of thousands of histone molecules synthesised in the cytoplasm enter the nucleus. Ribosome production in the nucleolus requires the continuous importation of great numbers of the approximately 80 different kinds of ribosomal proteins.

Experiments suggest that histones and other proteins may be actively transported through the pores. In addition, large numbers of mRNA-protein complexes and new ribosomal subunits flow out of the nucleus every minute. These substances are too large to be accommodated by the nine-nm opening of a pore, and some means of increasing the effective opening to facilitate their passage must occur.

The particle sometimes observed in the center of a nuclear pore may represent material caught in transit between nucleus and cytoplasm. During mitosis in most kinds of cells the nuclear envelope breaks into small fragments that disperse in the cytoplasm. These come together again to reform nuclear envelopes around daughter nuclei in telophase.

The specific reaggregation of these fragments shows that the membranes of the nuclear envelope must have properties that distinguish them from membranes of the endoplasmic reticulum. The re-formation of the nuclear envelope is closely coupled with the re-formation of the nuclear lamina on the nuclear surface of the inner membrane.

***Structure of condensed chromosomes***

As a cell enters prophase of mitosis or meiosis the interphase chromatin becomes tightly condensed by further packing, and individual chromosomes become recognisable. An early stage of such packing is represented by the lampbrush chromosome, in which the central axis has probably shortened and thickened compared to the interphase state. The central axis is, in turn, folded or coiled to produce the fully condensed chromatids of a chromosome. The condensation of chromosomes is achieved by proteins. The synthesis of a principal protein involved in condensation begins just around the start of prophase, but details about the events of condensation are still lacking.

Removal of histones and DNA from metaphase chromosomes leaves a residual structure that is still recognisable as a chromosome. The proteins that are left are believed to form the scaffold onto which the DNAhistone components are organised and packed into the condensed mitotic or meiotic form. The chromosome scaffold and the nuclear matrix of the interphase nucleus may well be forms of the same structure,

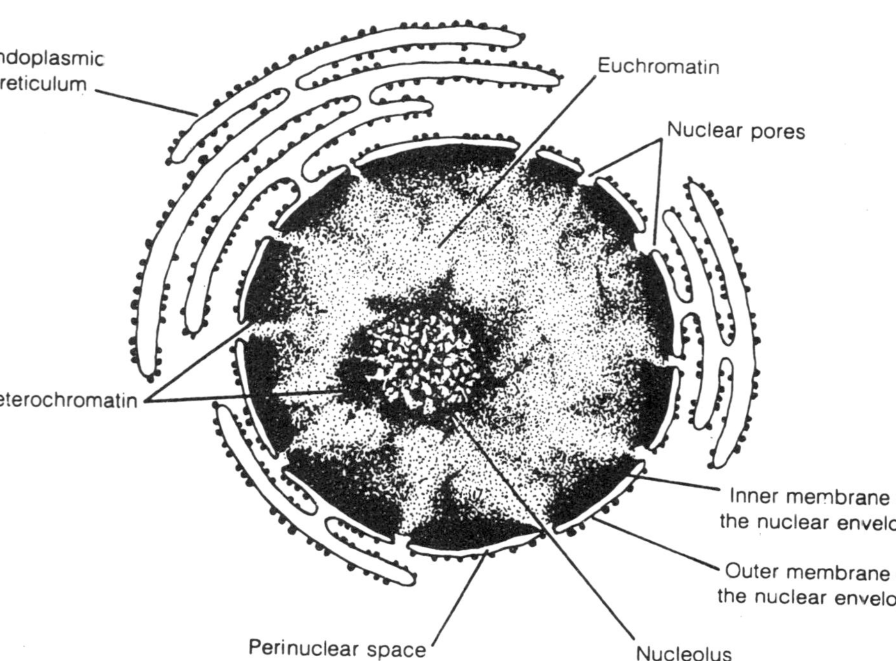


*Figure 8.26: Schematic diagram of a cross section of an animal cell nucleus. The nucleus is enclosed by an envelope composed of an inner and an outer membrane. The two membranes enclose the perinuclear space. The outer membrane has attached ribosomes and is continuous with the endoplasmic reticulum. The outer membrane connects to the inner membrane at nuclear pores. The nuclear face of the inner membrane is coated with the nuclear lamina. Heterochromatin is attached to the nuclear lamina and also surrounds the nucleolus.*

one designed to produce the compact mitotic chromosomes, the other responsible for the organisation of teh transcriptionally active interphase chromosome.

**Banding in Mitotic Chromosomes**

Mitotic chromosomes are not uniform in structure throughout their lengths; a considerable amount of substructure is revealed by a variety of staining techniques. The substructure takes the form of several kinds of transverse bands. Harsh treatments that extract DNA differentially from regions of a chromosome leave certain regions intact. In subsequent staining, these regions, known as C-bands, show up more intensely.

C-bands in general correspond to DNA that is packed as heterochromatin in the interphase nucleus. In metaphase chromosomes, this DNA appears to be more tightly packed than DNA that occurs as euchromatin during interphase. C-banding occurs predominantly in the centromeric region of mammalian chromosomes, where heterochromatin containing highly repetitive DNA is known to be located.

Three other staining techniques produce R-, Q-, and G-bands. The different patterns of C-, R-, Q-, and G-banding are constant in a species and allow each chromosome to be identified unequivocally Figure elsewhere in this book shows the banding map for the chromosomes of the human genome.

Banding in mitotic chromosomes is an entirely different phenomenon from the banding seen in polytene chromosomes. The latter are interphase chromosomes and their banding should not be confused with the banding in mitotic chromosomes. Banding patterns in mitotic chromosomes are useful because they allow segments of chromosomes that have been exchanged between chromosomes (*translocations*) to be precisely identified.

Translocations can be induced by radiation and sometimes they occur without known cause. They occur frequently in cancer cells, and often they have important consequences for the cell.

**Replication of Chromosomes**

In most eukaryotes the DNA molecules of chromosomes are enormously long, raising the question of how replication is organised. Prokaryotic chromosomes, which are shorter and circular, replicate beginning at a single origin; that is, they are single replication units. In contrast, the linear DNA molecule of a eukaryotic chromosome contains many points of initiation of replication; that is, it contains many replication units.

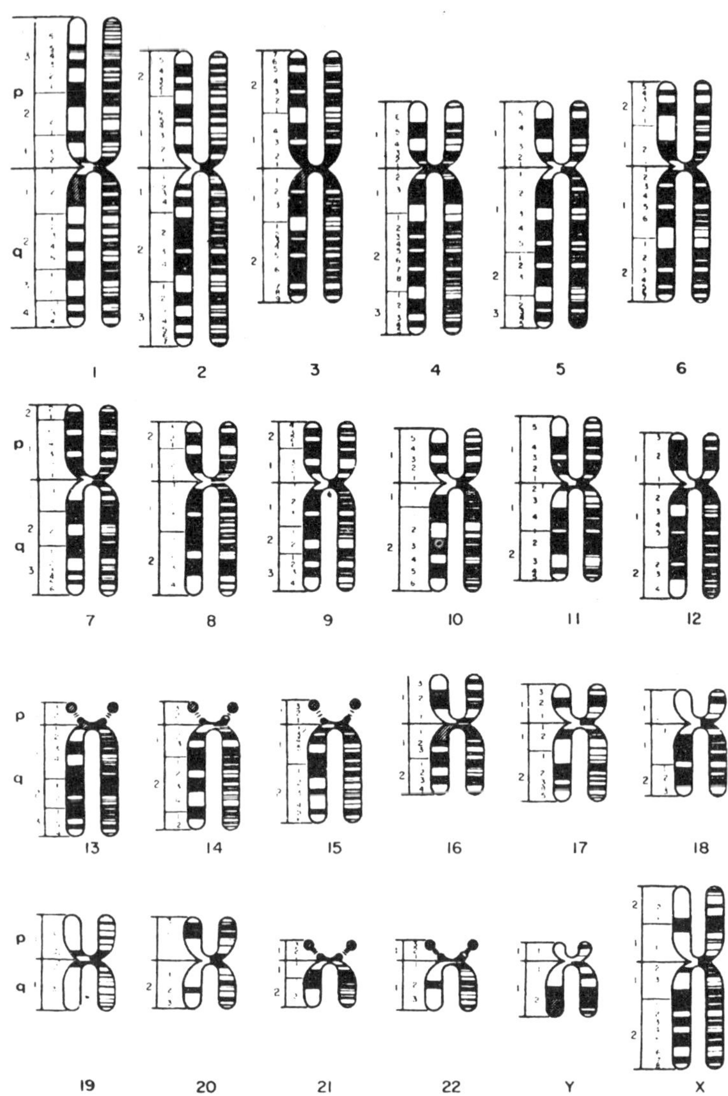


*Figure 8.27: A drawig of the 22 human autosomes and the X and Y chromosomes showing the bands that have been revealed by several kinds of staining. The crhomatid on the left in each chromosome shows the bands seen in metaphase chromosomes.*

### *Replication units*

The multiple replication units of eukaryotic chromosomes are readily seen by autoradiography or by electron microscopy. The autoradiographic experiment is simple. Cultured cells are incubated with $^3$H-thymidine for a few minutes during DNA replication. Cells are lysed with a drop of a solution of an ionic detergent on a microscope slide. The detergent lyses the nuclei and removes essentially all proteins from the DNA. The drop of lysed material is spread evenly over the entire microscope slide.

This draws the DNA molecules out in long, linear stretches. Subsequent autoradiography shows that DNA replication was occurring at many places along individual molecules during the short incubation with $^3$H-thymidine. Each stretch of radioactivity represents a replication unit. Replication can also be seen directly by electron microscopic observation of isolated DNA molecules. Each segment of the molecule that is doubled represents a separate replication unit.

In prokaryotic chromosomes origins of replication are defined by specific deoxynucleotide sequences. The strongest evidence that origins are similarly defined in eukaryotes comes from the study of yeast chromosomes. These contain ARSs discussed earlier in this chapter that confer on yeast plasmids the ability to replicate. ARSs that function as apparent specific origins of replication in yeast plasmids have been identified in the DNA molecules from a variety of eukaryotes. Conclusive evidence that these sequences function as origins of replication in the chromosomes of these eukaryotes is still lacking.

### *Bidirectional replication*

Replication is initiated at the single point within each replication unit and proceeds bidirectionally from that point much as in prokaryotes. Such points are labeled as origins in Figure elsewhere in this chapter, but whether they represent specific sequences has not been unequivocally proven. Bidirectional replication was discovered by the following autoradiographic experiment: fluorodeoxyuridine, which inhibits DNA replication, was used to block entry into the S period. A culture of cells was thus synchronised. Cells in the G2, M, and G1 periods of the cell cycle are unaffected by the drug, proceed to the G 1 /S transition point, and stop.

Cells already in S are killed by fluorodeoxyuridine. After enough time was allowed for cells to be synchronised at the G 1 /S border, the inhibitor was washed away and fresh nutrient medium containing $^3$H-thymidine was added. As a result, replication was initiated, with

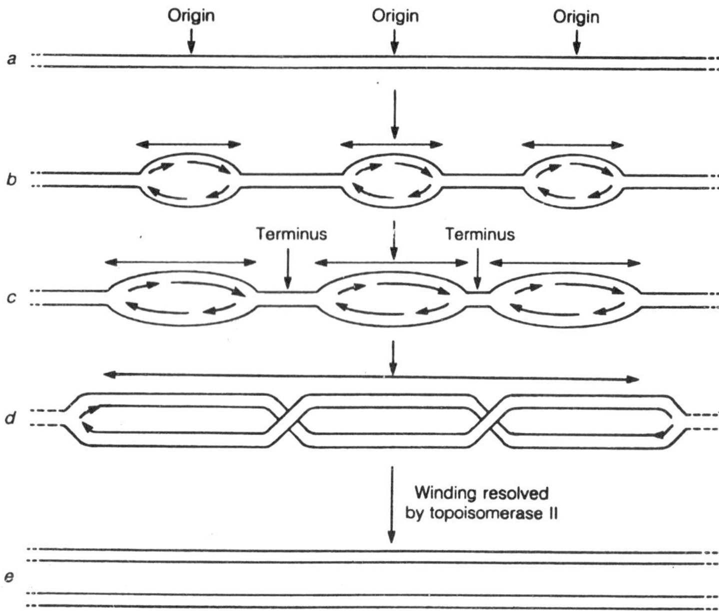


*Figure 8.28: Replication of successive replication units in a eukaryotic chromosome.*

incorporation of $^{3}$H-thymidine. After a few minutes of replication, nonradioactive thymidine was added to the medium. The thymidine was taken up by cells and phosphorylated to TTP, which became part of the pool of TTP used for DNA replication.

Over several minutes nonradioactive TTP replaced $^{3}$H-TTP in the pool of TTP in the cell. As a consequence the amount of radioactivity incorporated into DNA progressively diminished. Some minutes after addition of nonradioactive thymidine, cells were lysed; autoradiography of DNA molecules showed the expected tandem arrays of replication units.

However, each autoradiographic image was darkest in its central region and faded to imperceptibility at its two extremities. The fading images reflect the travel of replication forks during the decline of $^{3}$H-thymidine triphosphate in the cell and show that replication has proceeded bidirectionally from a single point (origin) in each replication unit.

***Termination between replication units***

Each replication fork eventually meets a replication fork traveling toward it in the next adjacent replication unit. Meeting of forks completes replication of'that particular segment of the chromosome, and the replication machinery dissociates from the DNA. The terminus between adjacent replication units is defined by the meeting place of replication forks rather than by a specific sequence in the DNA.

The meeting of forks and completion of replication between replication units leaves the two daughter molecules linked by one interloop between them for each replication unit. These interloops would be difficult to eliminate by untwisting of the two duplexes at the end of replication. An average of more than 1300 interloops are generated in a human chromosome. Loops are removed by the action of topoisomerase, which breaks the chains in the daughter double helices, allows the daughters to separate and then reseals the breaks. Mutant yeast cells lacking the proper topoisomerase cannot remove interloopings between the two daughter molecules. As a result, the two chromatids in a chromosome cannot separate from each other at the ensuing mitosis.

***Size and number of replication units***

An estimate of the size of replication units is obtained by measuring the distance between the centers of successive origins either in autoradiographic or electron microscope displays of DNA. Replication units in mammalian cells range from 15 to 100 $\mu$m (45,000 to 300,000 bp) with an average of 30 p.m (90,000 bp). The four cm of DNA in an average mammalian chromosome therefore contain over 1300 replication units. A haploid set of mammalian chromosomes contains about 33,000 replication units.

***Rate of replication***

The rate of travel of replication forks has been estimated by several methods, among them autoradiography. The simplest procedure is to synchronise cells by blocking them at the G I/ S border as previously described, then releasing the block and simultaneously adding $^{3}$H-thymidine. DNA isolated at intervals thereafter produces progressively longer autoradiographic images, which shows that in mammalian cells replication forks travel about one $\mu$m per minute (3000 bases per minute). Since the average size of a replication unit is 30 $\mu$m and each unit has two forks, the time to replicate an average-sized unit is 15 minutes.

## DNA MOLECULES IN MITOCHONDRIA

The mitochondria of all eukaryotes contain DNA molecules (mtDNA). These encode a few of the macro-molecules that make up mitochondria, although all but a few of the 100 or more different kinds of polypeptides in mitochondria are encoded by nuclear genes. The polypeptide products of these nuclear genes are synthesized in the cytoplasm and subsequently imported into mitochondria. Among protozoa, yeasts, and other unicellular eukaryotes mitochondrial DNA molecules range in size from 15,000 to 75,000 bp. In a few species the molecules are linear, but in most they are circular. There are multiple identical copies of mtDNA molecules in each mitochondrion, and all molecules in all the mitochondria have the same coding functions.

Among unicellular eukaryotes mtDNA of the budding yeast has been studied intensively. It is a circular molecule containing 75,000 bp, and there are ten to 50 DNA molecules altogether in all the mitochondria of a single cell. These mtDNA molecules represent 5 to 25 percent of the total cellular DNA but only 0.5 percent of the total sequence complexity of the total DNA.

These are unusually high percentages, and are due to the unusually large size of yeast mtDNA (75,000 bp) and the small size of the yeast nuclear genome ($1.5 \times 10^7$ bp). In all other eukaryotes studied so far, mtDNA represents a far smaller percentage. In mammalian cells mitochondrial DNA makes up less than one percent of the total cellular DNA and about 0.0005 percent of sequence complexity of the total DNA in the cell.

Molecules of mtDNA in invertebrate and vertebrate animals are closed circles as in yeast but are much .smaller, ranging in size from 13,000 to 18,000 bp. In mammalian mitochondria, mtDNA molecules contain about 16,000 bp.

The entire sequence of 16,569 by in human DNA has been determined. There are five to ten molecules per mitochondrion and 100 to several thousand mitochondria per cell, depending on the size and type of cell. Fibroblasts have about 100, liver cells about 1000, and heart muscle cells many thousands.

In plants, mtDNA molecules are much larger and more variable in size than those in animals. Even within a single organism the lengths of mtDNA molecules are variable, and some molecules are circular and some are linear. These variations are apparently not the result of DNA breakage during preparation and are not understood. Much less is known about the coding functions of mtDNA in plants compared with yeast and animal mtDNA, but evidence so far suggests that plant mtDNA contains about the same number of different kinds of genes as other eukaryotes.

Mitochondrial DNA molecules are not complexed with nuclear-type histones to form nucleosomes. Histone-like molecules are present instead, and the mtDNA-protein complexes form nucleoid bodies in the matrix compartment. Mitochondria reproduce by fission, and the multiple mtDNA molecules, which are located in the matrix of mitochondria, are distributed randomly between daughter mitochondria. In some kinds of cells, for example, chick cells in culture, and probably in cells in general, mitochondria fuse and separate often, so that the population of mtDNA molecules are dynamically shared among the mitochondria within a single cell.

A yeast mtDNA molecule is five times longer than a mammalian mtDNA molecule but both encode almost an identical set of genes. Most of yeast mtDNA has no known coding or other function. MtDNA molecules in both organisms encode three of the subunit polypeptides of cytochrome c and seven or eight other polypeptides. Both encode the two major rRNA molecules for large and small subunits of ribosomes. These rRNAs are smaller than those encoded by nuclear genes, and mitochordrial ribosomes are smaller than cytoplasmic ribosomes. Mammalian mtDNA encodes 22 kinds of tRNA molecules and yeast encodes about 25.

Thus, mtDNA encodes some of the molecules needed to make up the translation machinery in the mitochondrial matrix. However, most elements, like most ribosomal protein and aminoacyl-tRNA synthetases, are encoded by nuclear genes, synthesized in the cytoplasm, and imported into the mitochondrion. The mitochondrial translation machinery is limited in use to the several kinds of mRNA molecules transcribed from mtDNA. No mRNA molecules are imported from the cytoplasm. Like bacterial mRNA molecules, mitochondrial mRNAs are not capped at their 5' ends. They have short, 3'-poly(A) tails.

Transcription of human mtDNA is unusual. One long transcript is produced from *each* strand of the DNA, in contrast to transcription from only *one* strand in nuclear DNA. The two transcripts are full-length copies of the two DNA strands. Both transcripts are processed, each yielding some of the rRNA, tRNA, and mRNA products. The coding segments for some genes are in one strand of the DNA and coding segments for others are in the opposite DNA strand.

Remarkably, the genetic code of mtDNA varies slightly from the nearly universal code present in the nucleus of eukaryotes, the nucleoid of prokaryotes, and viruses. In fungi, which includes yeast, and in animal cells, the codon UGA functions as a codon for tryptophan rather than as a stop codon for translation. In the mtDNA of some organisms AUA codes for methionine rather than for isoleucine. AGA and AGG code for stop rather than arginine.

During the cell cycle the total number of mtDNA molecules doubles, keeping pace with a doubling in mitochondrial number. Replication is catalysed by a mitochondrial-specific polymerase called DNA polymerase y, which is encoded by a nuclear gene. The mtDNA molecules replicate asynchronously with each other. Some molecules may replicate twice during a given cell cycle and others not at all. The time of replication is spread over the entire interphase and bears no relationship to the nuclear S period.

At least some segments of mtDNA are represented by nearly identical sequences at scattered locations in the nuclear genome. For example, the 16S rRNA sequence of human mtDNA is also present in nuclear DNA. The significance of mitochondrial sequences in the nucleus is not known.

## DNA MOLECULES IN CHLOROPLASTS

Chloroplast DNA molecules (ch1DNA) range in size from 33,000 by to more than $2 \times 10^6$ by. In higher plants ch1DNA molecules consist of about $1.5 \times 10^5$ bp. The number of molecules per chloroplast ranges from 20 to more than 100, depending on the species, and these are distributed in multiple nucleoid-like bodies in the stromal compartment of the chloroplast.

More than 80 polypeptides are encoded by ch1DNA, as well as rRNAs for large and small ribosomal subunits and tRNAs. Many other chloroplast proteins are encoded by nuclear genes, are synthesized in the cytoplasm, and are imported into the chloroplast. For example, the small subunit polypeptide of the enzyme ribulose-l,5-bisphosphate carboxylase is encoded by a nuclear gene; the large subunit polypeptide is encoded by ch1DNA.

This enzyme occupies a pivotal position in the conversion of $CO_2$ to carbohydrate during photosynthesis. Messenger RNA molecules transcribed from ch1DNA are translated by the chloroplast translation machinery in the stromal compartment of chloroplasts.

Like mtDNA molecules ch1DNA molecules double in number during the cell cycle and are distributed randomly between the daughters during chloroplast fission. Histones are not associated with chlDNA.

# 9

# CELL DIVISION

Cells that are growing are destined to divide. Cell division takes place at different intervals which depend upon the species and the environmental conditions. As already mentioned, *Escherichia coil"* on a medium containing water, glucose, ammonium chloride, potassium basic phosphate, sodium sulfate and trace metals, kept at 37°C., divides every 45 minutes. If supplied with amino acids and some other organic materials at 37°C., it divides every 20 minutes. Eukaryotic cells divide at a slower rate than bacteria.

It is true that eggs of some marine animals from warm climates divide about every 30 minutes, and those from cooler climates divide at about one hour intervals. However, such cells have reserves and need only convert these to cell constituents before dividing. Protozoan cells divide somewhat more slowly; *Tetrahymena pyriformis,* under favourable conditions, divides every 3 to 4 hours, and *Paramecium aurelia,* every 6 hours. Cells of higher plants and animals divide about once or twice a day. Thus the generation time (time between divisions) for cells of the broad bean *Vicia faba is* about 14.3 hours; for the cells of *Tradescantia paludosa,* about 17 hours; for Chinese hamster fibroblasts, 11 hours; and for mouse fibroblasts, 22 hours. Cell division or *mitosis,* which consists of prophase, metaphase, anaphase and telophase, occupies only a small part of the generation time, as may be seen in figure elsewhere in this chapter. The major portion of the division interval is interphase.

Interphase consists of three main stages: $G_1$ (Gap 1) from telophase to the beginning of S (synthesis); S, during which DNA synthesis takes place; and $G_2$ (Gap 2), from the end of DNA synthesis to the beginning of mitosis (division). The amount of time for DNA synthesis varies, as seen in figure elsewhere in this chapter. It is evident that in a multicellular plant or animal cell division goes on until the adult stage is reached. Thereafter, cell division occurs only in germinal tissues such as the cambium in the stem and the root of a plant and the germinal layer of the epidermis, the gametogenic cells in the gonads and blood cell-forming tissues of the animal. Divisions also occur during repair of wounds in both plants and animals.

## SYNCHRONISED CELL DIVISION

For many of the studies of cell division it has been found desirable to have all of a population of cells in a culture divide at one time. Such *synchronised* cell division makes possible more effective analysis of the various components of the process. Eggs of marine animals have been

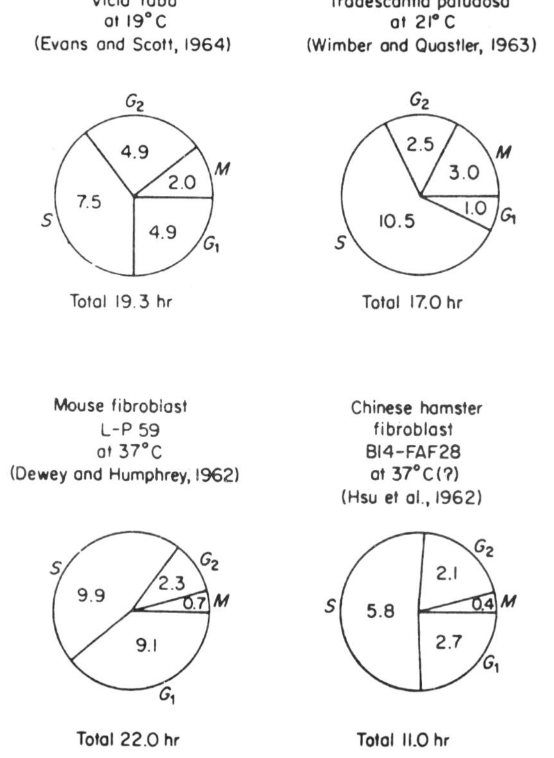


*Figure 9.1: Duration of phases in the division cycle of a cell: $G_1$ (Gap 1), S, (synthetic phase), $G_2$ (Gap 2) and M (mitosis). $G_1$ follows mitosis and lasts until the beginning of DNA synthesis ($S_1$); $S_1$ lasts until replication of DNA is over, when $G_2$ begins and lasts until the beginning of mitosis.*

favourite objects for such studies since practically all the eggs divide at the same time in a suspension of healthy and normal cells.

This synchrony in dividing marine eggs implies that some event has occurred to achieve the synchrony, e.g., the almost simultaneous entry of sperm in all eggs starts a train of events that takes about the same amount of time in each of the cells. Cleavage in the embryo continues in synchrony for several generations unless cells are separated from one another. When the eggs have been affected by unfavourable conditions, e.g., temperature, radiations, pH, division may be delayed, and when it starts, some eggs are found to cleave before others.

The susceptibility of the cells to the unfavourable conditions varies about a mean for the population and a characteristic distribution curve is obtained for onset of division in the population. In a culture of bacteria, protozoa or tissue culture cells the number of cells in division at any one time is limited, being from 5 to 10 per cent. The ratio of the number of cells in division to the total number of cells is called the *mitotic index*.

The mitotic index appears to be a function of the *generation time* for a given species under the conditions provided and the actual length of time a cell of this species remains in mitosis. The mitotic index may be increased by selecting a single cell from such an asynchronous culture and using it as the progenitor of a culture. For a few generations

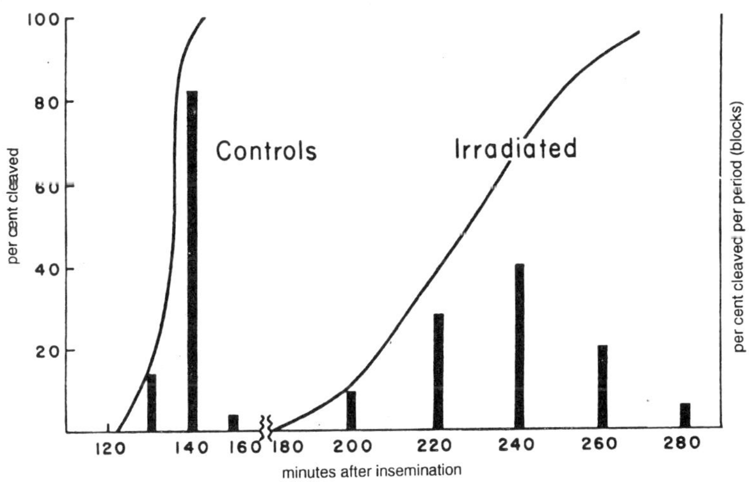


*Figure 9.2: High degree of synchronyu in division of normal sea urchin eggs (inseminated simultaneously) as shown in the curve for "controls".*

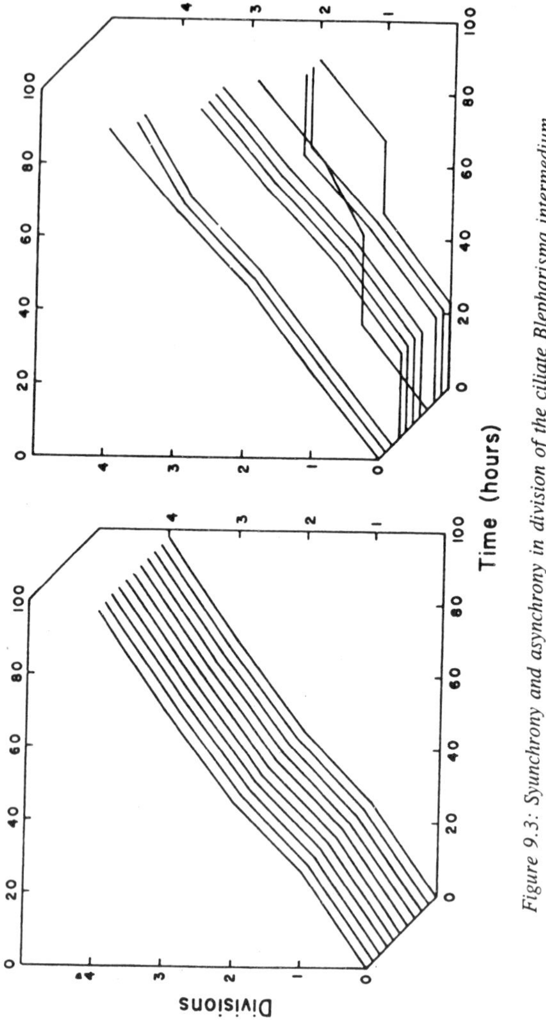


*Figure 9.3: Syunchrony and asynchrony in division of the ciliate Blepharisma intermedium.*

thereafter, synchrony is obtained, but it also gradually disappears. A variation of this technique consists of filtering bacteria of a given size from a mixed population of cells, thus obtaining cells in essentially the same stage of growth which will divide at approximately the same time.

The failure to achieve synchrony of division in a culture of cells is probably a result, in part, of differential exposure to various environmental conditions. In a syncytium (multinucleate cell) in which mixing of all materials occurs and the conditions of the environment are similar, the division of nuclei is usually perfectly synchronised. Also, in the syncytial insect egg nuclear division continues synchronously for as many as eleven generations.

The same is true for the nuclei in the syncytial endosperm of plant embryos. It is thought likely that nuclei in division may secrete division-stimulating materials. To explain the low mitotic index in most cell cultures it has been suggested by some that the division-stimulating material from the few cells in division does not affect many other cells, either because they are distant or because the membranes of such cells are relatively impervious to the hypothetical division-controlling substances.

This concept of control of synchrony by secretion of materials from nuclei has some experimental backing. For example, grafts can be made between two multinucleate amebas *(Chaos chaos),* each with nuclei synchronously dividing but out of phase with one another. After one cycle the division of all nuclei from both amebas now present in one cell is synchronous, the larger piece imposing its time upon the smaller. There are many examples of synchronised cell division which can be found in cells of higher plants and animals. Cells in synchronous division are found, for example, in spermatogenic cells lining the lumen of a germinative tubule, which are connected to one another by cytoplasmic bridges. Such bridges insure entry of the division-stimulating material from one cell to another.

It is important at this time to emphasize that a suspension of cells, whether they are cells in tissue culture, protozoans, yeast or bacteria, are fundamentally different from a suspension of marine eggs. Marine eggs have undergone a period of growth in the ovary and they have self-contained supplies which will serve for many cell divisions before intake of nutrients need take place.

Cells other than eggs, however, must incorporate nutrients and grow before they can divide. Perhaps not all cells are able to incorporate the same nutrients at the same - rate. Studies on *Tetrahymena* (ciliate protozoan) show that the mass of the cell increases linearly as measured by oxygen consumption, by increase in volume and by $^{14}C$-methionine incorporation, during the entire *synthetic period, (S),* from the onset of furrowing in one division until about 10 to 20 minutes before the next division. The division cycle has a generation time of 2.25 to 2.5 hours at 28 to 29°C.

Furthermore, $^{3}H$-histidine incorporation (protein synthesis) and $^{3}H$-uridine incorporation (RNA synthesis) continue during division with little change. During the 10 to 20 minute *predivision period,* which precedes furrowing, there is no increase in mass while the cell gets ready to divide. Such a period has also been found in other cells; e.g., in *Amoeba* it occupies about one-sixth of the generation time.

Studies on *Paramecium aurelia* show that the mass, as measured by increase in dry weight, changes little for a period after division, and then it increases almost exponentially. From the data recorded this would appear to be exceptional compared to most cells studied. On the other hand, Lovlie in a careful study on single *Tetrahymena* (using the diver technique) has shown that the type of curve obtained for increase in mass with time is related to the conditions of growth. He found three types of curves for *Tetrahymena,* exponential, linear and linear with a plateau, depending on whether the growth was balanced (doubling during the interdivision period) or unbalanced.

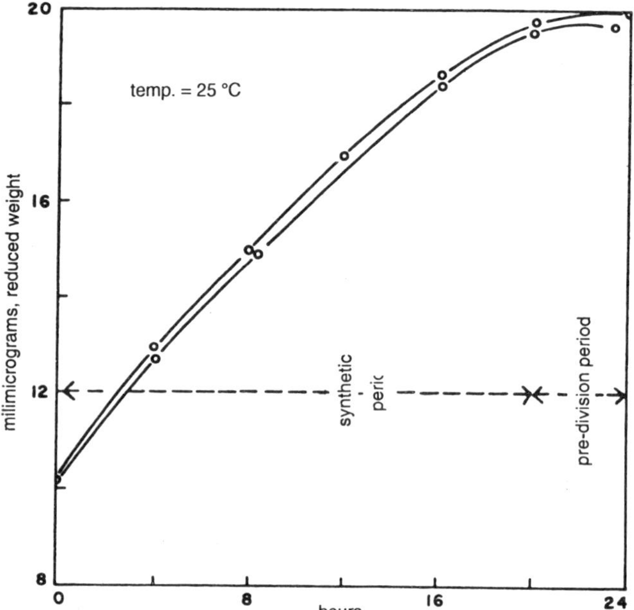


*Figure 9.4: Growth in mass of* Amoeba proteus. *The pair of lines represents the growth of a pair of sister cells from the time of division of their mother cell to the time they themselves divide.*

Since it is not possible to achieve synchrony of cell division in cell suspensions other than marine eggs, attempts have been made to *shock* cells into synchrony. Two methods have been used: chemical shock and physical shock. Chemical shock consists of withholding or limiting the supply of some nutrient necessary for division and then supplying it to the culture at one time in a large quantity, inducing in this manner a high level of simultaneous metabolic activity. The physical shock is one that is unfavourable for the act of cell 'division yet favourable to other metabolic activities preceding division, thus allowing the cells in the predivision stages of the division cycle to catch up with those in the later stages of the division cycle.

Only a few experiments using shocks for synchronisation of cell divisions will be considered here. Cell division and DNA synthesis in *Escherichia coli T-15,* a thymine-requiring mutant (thymineless), are blocked immediately upon transfer of the organism to a thymine-free medium. However, RNA and protein synthesis continue, apparently at the original rate. When thymine is added 30 minutes later, DNA synthesis is resumed and nearly all the cells are found to divide simutaneously after a lag of 35 to 40 minutes. Similarly, in *Lactobacillus acidophilus,* synchrony can be induced by the addition of thymidine to a thymidine-starved culture.

Yeast cells starved in succinate buffer until some of the reserves are gone will divide synchronously after return to complete nutrient medium, including carbohydrate and nitrogen sources. Similar results were obtained with a variety of cells needing some particular metabolite. For example, synchronous division in the cells of the epidermis of the insect *Rhodnius follows* its periodic ingestion of blood. In *Chlorella* and various algae, all photosynthetic, the daily periodic lighting regimen makes possible synchronisation of cell division, presumably by periodic accumulation of food reserves during the period of illumination. Lighting, too, may be considered a physical shock, much like temperature; in fact, all the arguments used for temperature apply to light in light-sensitive cells. Environmental changes produce oscillations in growth of many types of cells.

Use of temperature as a physical shock to obtain synchronised division stems from the notion that the processes that occur during the division cycle are differentially sensitive to temperature. If some reaction in the predivision period is more sensitive to heat than are the reactions in the synthetic period, then a high temperature might prevent division without stopping syntheses.

Thus, cells lagging behind in preparations for division might be given a chance to catch up with the others. As expected, a single temperature shock synchronises only a small fraction of the cell population because it allows only a small proportion of the cells to accumulate in

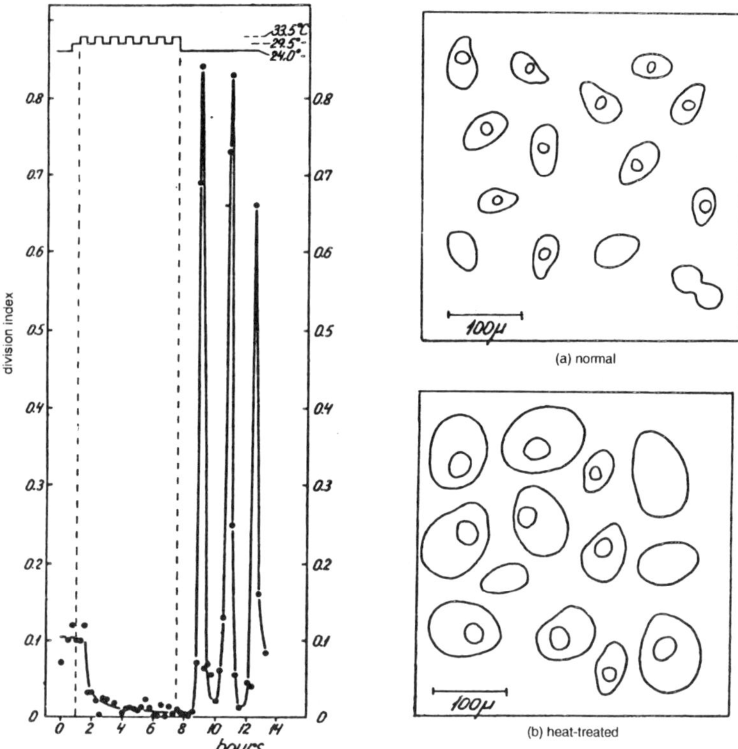


*Figure 9.5: Left, Synchronisation of division by heat-shocking Tetrahymena. Seven shocks at 33.5°C. are given in succession, alternating with 29.5° C. At the conclusion of the heat shocks, the temperature is dropped to 24.0°C. The three post heat treatment divisions are synchronised to the extent indicated by the division index. Right (a), Normal Tetrahymena drawn with camera lucida and (b) Tetrahymena after heat treatment. The circles inside the cell are the macronuclei.*

predivision stages. On the other hand, a series of temperature shocks which block division-each alternating with an exposure to a near optimal temperature for a period insufficient to allow the cells in the predivision stages to divide -synchronises most of the cell population, because of the gradual accumulation of a large proportion of the cells in predivision stages before cessation of heat shocks.

Zeuthen and Scherbaum found that alternate exposure of suspensions of the ciliate *Tetrahymena,* for half-hour periods at 28 to 29° C. (optimal) and 34° C. (inhibitory) for seven cycles resulted in 85 per cent synchronisation of the cells in division when the cultures were subsequently kept at 28° C. Synchrony persisted for several cell generations and then division became random. Thermal shocks have also been effective for synchronising division of many other kinds of cells, including bacteria. Cold shocks have also been used to induce synchronisation of cell division in a number of protozoans and bacteria. Nutritional deficiency, coupled with temperature shocks, has also been very effective. X-rays have also been shown to induce synchrony in division, again presumably by holding back the cells in the radiation-sensitive predivision stages and causing accumulation of the cells which will ultimately divide at nearly the same time.

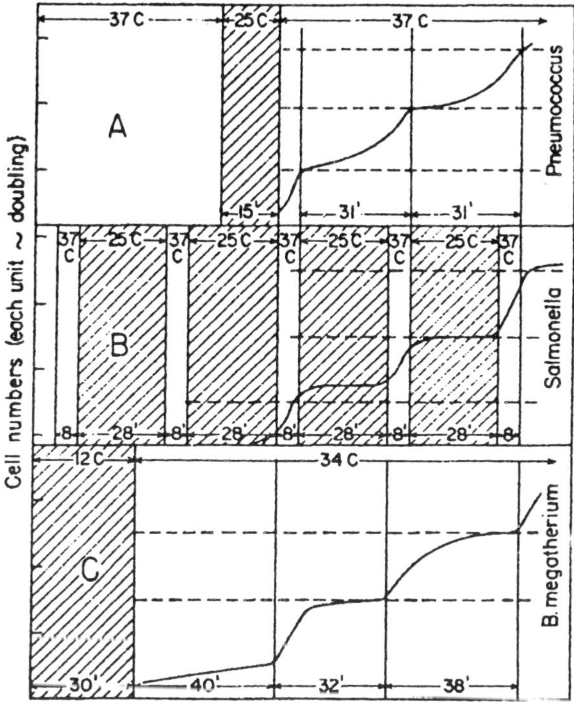


*Figure 9.6: Three different examples of synchronisation in Bacteria. A, Pneumococcus. B, Salmonella typhimurium . C, Bacillus megatherium.*

Much effort has gone into production of synchronised cultures of cells because in such cultures division and growth are essentially uncoupled. Therefore, growth can be studied in the cells prior to division, and the process of division and its accompaniments can then be studied in the cells accumulated in the predivision stages. The cells shocked into synchrony have accumulated sufficient reserves for a series of divisions, which follow one another more rapidly than the usual generation time for the species until the accumulated reserves have been partitioned among the descendants. This, of course, is also true of a suspension of

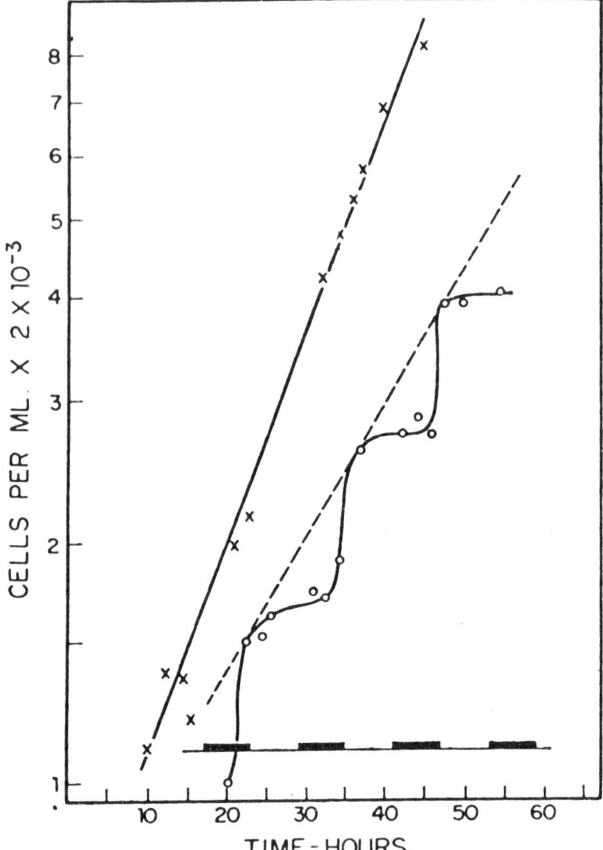


*Figure 9.7: Increase in cell population of the photosynthetic dinoflagellate Gonyaulax polyhedra in continuous light (X) and in alternating light and dark periods of 12 hours (open circles). Light intensity, 900 foot-candles; temperature, 21.5° C.*

marine eggs. Heat-shocked cells thus grow until they are considerably larger than controls. In a heat-shocked culture of *Tetrahymena*, the generation time for the first few synchronised divisions is about 60 per cent of that required for a culture that was not subjected to heat shock.

Division of heat-shocked *Tetrahymena* continues even in the presence of inhibitors of synthesis of DNA and steroids. However, certain long chain unsaturated fatty acids must be synthesized, as must certain specific proteins. The specific protein synthesis continues even if the cells are starved.

The specific protein is presumably "division protein" thought to be required for structuring the cell for division. It has been extracted and characterised. Interestingly, DNA synthesis, as measured by $^{3}$H-thymidine incorporation, is not synchronised with cell division in heat-shocked cultures. It would be of interest to determine the DNA polymerase activity at various times during the cell division cycle.

Synchronised cultures of various cells in suspensions are now being widely used in biochemical and cytochemical research. It is important to note, however, that changes in size and composition of the cells after synchronisation must be taken into consideration. For example, resistance of such cells to ultraviolet radiations is markedly altered.

Synchronisation of cell division in the photosynthetic dinoflagellate *Gonyaulax*, in the normal diurnal rhythm of day and night, occurs in 85 per cent of all cells destined to divide during a 5 hour period spanning the end of darkness and the beginning of the light period. Although it might appear that cell division here, as in *Chlorella*, is related to periodic changes in nutritional conditions, keeping *Gonyaulax* in continuous dim light only sufficient to maintain nutritional balance in the cells does not stop synchronous division for at least 14 days.

The synchronous division here is also relatively independent of small variations in light intensity and temperature. In high light intensity, however, synchrony is lost in 4 to 6 days. This is an interesting case worthy of detailed study. At present it is interpreted as an example of inherent biological rhythm.

Bruce lists the generation time of cells from a large number of species of plants and animals. In many cases the division of cells appears to be circadian, almost every 24 hours, but it has not yet been determined whether the day-night cycle will synchronise these to a 24 hour rhythm.

## SOURCE OF ENERGY FOR CELL DIVISION

Cell division, as has been pointed out many times throughout the text, is a fundamental activity of the living cell. During cell division the cell does work at the expense of energy derived from nutrients, as witness the case just cited of synchronised cell division in yeast obtained whenever the cells are supplied with adequate food. It has also been shown that addition of glucose to isolated epidermal tissue culture results in synchronous division of many cells. Presumably, cell division had previously been blocked by lack of nutrient since such cells store little glycogen. Even if glucose is supplied, division fails if oxygen is not available.

Glucose may be replaced by lactate, glutamate, fumarate or citrate.

All the experiments suggest that operation of the Krebs cycle supplies the energy for division of the cells. As expected, mitosis is inhibited by Krebs' cycle poisons such as malonate, cyanide and carbon monoxide, and by phlorhizin, a phosphorylation inhibitor.

The latter finding suggests that high energy phosphate bonds are an energy source or are involved in building this source. Marine eggs and protozoans require oxygen for division, but frog eggs and many embryonic tissue cells do not, presumably supplying their energy for cell division by glycolysis. Cells which require oxygen for division have a lower rate of glycolysis compared to their rate of respiration, Cells which divide in absence of oxygen have rates of glycolysis much higher than that of respiration, sometimes severalfold greater. As might be expected, in cells which can supply their energy needs for division by glycolysis cell division is very sensitive to glycolytic inhibitors such as iodoacetic acid. In order to have an effect on cell division the energy sources must be supplied early in the cell division cycle during what Bullough calls the *antephase*.

Once a critical concentration of the energy-rich substances has been built up and the prophase begins, he found that nothing short of killing the cell will stop it from dividing. This has been noticed not only for cells deprived of oxygen or metabolically poisoned, but also for those damaged by radiations or subjected to other injuries; such cells divide and soon thereafter cytolyse. Initiation of mitosis appears to begin a series of concatenated and irreversible reactions which stop only when the cells have divided.

By employing carbon monoxide, working under green light, to block cytochrome oxidase in sea urchin eggs and releasing the block by shifting from green light (which was not thought to be absorbed by the enzyme-carbon monoxide complex) to white light, Swann obtained the following results. If the inhibitor is applied before sea urchin eggs enter mitosis, the first cleavage is delayed by a time about equal to the time of application of carbon monoxide.

If the inhibitor is applied after the cells have entered mitosis, they complete mitosis and cleave with little or no delay, but the second cleavage is delayed for a period which is roughly equal to the period of inhibition. It was suggested that these results can best be explained by a hypothetical *energy reservoir* which is continually being filled and in which energy is stored for a cell division during the preceding mitosis and cleavage. Once the energy has siphoned out it carries the cell through mitosis and cleavage; if at this time the cell is poisoned for a period by monoxide and energy is no longer being accumulated in the reservoir, cleavage will nevertheless continue to completion. However, the next cell division is delayed for an interval equal to the period of application of the poison. Lack of oxygen and various other poisons that affect the aerobic enzymes would also affect cell division in a similar manner.

Poisons which do not affect energy-liberating systems act in a different manner. For example, ether (1 per cent) is almost without effect if applied to a suspension of sea urchin eggs before the mitotic spindle has formed, but if applied after that time, it blocks development of the spindle fibers, maintaining the proteins in a solvated state. If the treated eggs are then washed free of the ether, development proceeds at once, the only delay being the period during which the eggs were subjected :o the ether since in this case the energy reservoir, if it exists, is unaffected by ether.

The need for a postulated *energy reservoir* is difficult to accept on the basis of data collected both on animal and plant cells. Epel found that cleavage rate in the eggs of the purple sea urchin was decreased when the concentration of carbon monoxide was at a level which inhibited respiration. At this time the ATP level was also decreased. When the ATP level had dropped to 50 per cent of normal, mitosis was completely inhibited.

By varying the carbon monoxide concentration, and thus obtaining varying degrees of inhibition of ATP production, Epel demonstrated that the mitotic rate paralleled the ATP level in cells. He also found that division could be blocked at any stage of mitosis if the inhibitor was applied at the appropriate time. When phosphorylation is uncoupled by dinitrophenol (DNP), ATP fails to accumulate and cell division is also blocked. Such a concentration of DNP, which uncouples phosphorylation, causes an increased rate of respiration, but the respiration is now useless and idling.

This points again at ATP as the likely source of energy for cell division. The discrepancy between the results of Epel and those of Swann perhaps rests upon a difference in experimental technique. Swann did his "dark" experiments in green light, which is theoretically not effective in reversal of carbon monoxide poisoning; he used white light for full reversal of carbon monoxide poisoning.

Epel, on the other hand, put his carbon monoxide experiments in full darkness because he found that in green light the rate of respiration and ATP production were reduced to only 75 per cent of the level obtained by the same concentration of carbon monoxide in full darkness. Heat shock, as shown earlier, does not stop growth although it prevents division, and, consequently, the cells may reach a size four times that of the controls. Returning the cells to the optimal temperature after one or more heat shocks permits cleavage in less than the generation time.

Presumably the mitogenic and growth channels are separate at their definitive ends, though both are using the same sources of energy and materials. When the mitogenic channel is blocked by physical or chemical means, the energy and material which might normally go through this channel pass instead into synthetic reactions leading to extra cell growth. Consequently, after removal of the block, some time must elapse before the specific molecules necessary for cell division are synthesized to the necessary level.

In *Tetrahymena* if heat shocks as used by Zeuthen continue well beyond seven cycles, the cells may ultimately divide, even during a heat shock. This is taken to mean that accumulation of products of growth in a cell, and especially of the material needed for cell division has reached a point at which division is initiated. Division can then no longer be prevented by thermal shocks that previously blocked it.

Perhaps an even more clear-cut separation of growth processes from cell division is shown in experiments with *Tetrahymena* cells which are heat shocked in nutrient medium and then transferred to balanced salt medium with no nutrients. Such cells subsequently divide at least twice in absence of nutrients. Cleavage of such cells under these conditions can be "set back" or delayed by various metabolic poisons or physical shocks.

Experiments with dinitrophenol also distinguish two phases of metabolism in *Tetrahymena*. One, constituting about 30 per cent of total metabolism, relies on endogenous reserves and can support cell division; the other phase (about 70 per cent of total metabolism) depends upon exogenous supples and is coupled with growth.

It has become increasingly evident that specific protein *(division protein)* probably plays an important role in cell division in *Tetrahymena*. It is thought that a critical concentration of this particular protein is needed for cell division. It has also been shown that ATP, GTP and RNA accumulate in the period just preceding cytoplasmic division and that the changes are reversed at mitosis.

However, such an accumulation of free energy sources is interpretable on the basis of protein synthesis, because protein synthesis decreases for a period beginning just about two thirds of the time to the next cell division. At this time the protein formed on the ribosomes is less readily released than it was previously. The call for ATP is then less than its concentration in the cell, along with that of GTP, and RNA therefore increases. This change in concentration thus appears to be incidental to the slowing or cessation of protein synthesis rather than due to accumulation of an energy reservoir. The high energy phosphates then play a part in reversing the binding of the protein to the ribosomes.

## BLOCKING CELL DIVISION

### By Radiations

Blocking of cell division by x-radiation was observed soon after the discovery of x-rays, and this led to the use of ionising radiations in the treatment of cancer. Almost without exception dividing cells are most sensitive to x-rays during a short period just prior to the onset of mitosis. Cells already well along in the division cycle complete the division, but those in earlier stages regress.

The type of cell seems of little consequence provided the cell is growing and dividing either in culture or in the living organism. (The cells in a mature tissue are apparently radioresistant, i.e., not killed, when they have ceased dividing.) As a consequence, as already mentioned, x-rays produce a degree of synchronous division of cells in tissue culture. The locus of action of the x-rays is not fully known.

The rate of incorporation of precursors into DNA is decreased or inhibited in cells of tissues subjected to x-rays. Since in most cases assays were made many hours or even days after the irradiation, at which time the cell population may have declined as a result of x-raying, the results from this type of experiment have been questioned. However, by using cell suspensions, it has been shown that application of a certain x-ray dose, at a time just before synthesis of DNA has begun, is much more effective in stopping DNA synthesis than if the radiations are applied after synthesis has started. There thus appears to be a system connected to, but not identical with, DNA synthesis that is even more sensitive to x-rays than DNA synthesis itself. Larger doses affect both this process and synthesis of DNA as well.

It is interesting that synthesis of protein and RNA continue in *a radiationsterilised* cell (a cell that will not divide again). The result is the production of giant cells. Unbalanced growth of this type is followed by death, but the undivided cell may live for a considerable time. During this time, various materials leak from these cells which are found useful in experiments with tissue culture cells. For example, Puck has used mats of similar nondividing (sterilised) cells on which to place a single normal unirradiated human tissue cell. The cell lives using leaking material as nutrient and gives rise to a colony, the cells of which form a *clone;* that is, they are all the descendants of the single cell placed there.

Ultraviolet radiations also retard or stop cell division, but, because the radiations are absorbed so superficially, they are not suitable for cancer therapy. The action spectrum for cell division-inhibition in a suspension of cells resembles the absorption spectrum of nucleic acids or nucleoproteins. Also, it has been shown that DNA synthesis is retarded by small doses of ultraviolet radiations.

As in the case of x-rays, exceedingly small doses may inhibit division of cells without having any detectable effect on DNA synthesis, indicating the existence of a process even more sensitive than DNA synthesis but probably closely related to it. The action spectrum for the effect implicates nucleic acid.

In general, synthesis of DNA appears more sensitive to ultraviolet radiations than is RNA synthesis or protein synthesis. Since the blockage of DNA synthesis leads to unbalanced growth and larger doses of ultraviolet radiation inhibit RNA and protein syntheses, the interrelations between all these syntheses are complex.

**By Poisons**

A great number of poisons block cell division, but only those which help identify metabolic reactions or events in the complex which makes up cell division can be mentioned here. The action of such poisons has been studied extensively because of its theoretical interest and its possible use in control of cancer.

Many mitotic poisons interfere with interphase growth. For example, *metabolic poisons,* such as the inhibitors of respiration and glycolysis or those which uncouple phosphorylations from energy-yielding reactions, prevent the accumulation of high energy compounds and prevent release of energy for synthesis or division. Another group of mitotic poisons-the *antimetabolites-interfere* with syntheses.

Still others, such as the *antibiotic chloramphenicol* (isolated from *Streptomyces*), which resembles phenylala-nine and presumably competes with the latter on the surface of the protein-synthesising enzymes, block protein synthesis. If applied in sufficient concentration early in the division cycle, chloramphenicol also blocks cell division. *Tetracyclines* (isolated from *Streptomyces)* and *puromycin* interfere with protein synthesis, although probably in some manner other than that of chloram phenicol.

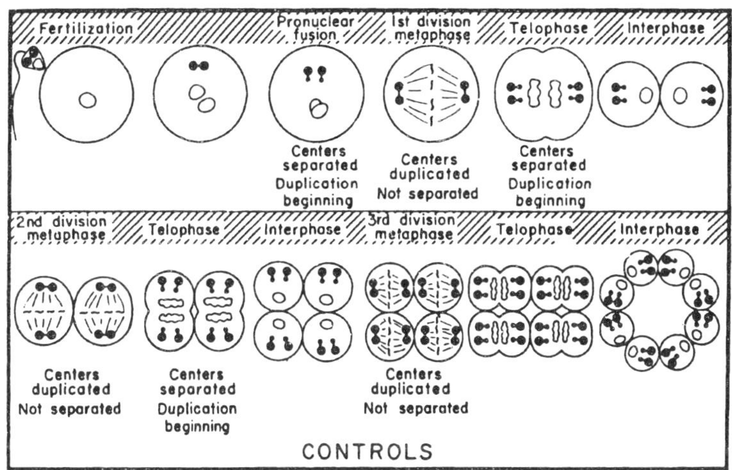


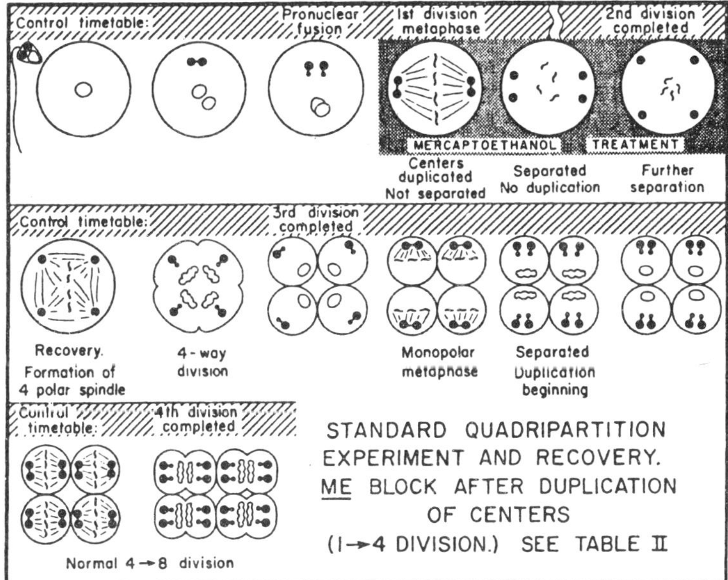


*Figure 9.8: A Diagrammatic interpretation of the reproductive cycle of the cell centres. All nuclear figures shown are purely symbolic and are introduced merely to identify mitotic stages.*

A few mitotic poisons block nucleic acid synthesis. These compounds resemble some normal metabolite, and thus compete with the normal metabolite for the surface of the particular cell enzyme (competitive inhibition). For example, *azaguanine,* which resembles the 'purine guanine, acts in this manner, and many other purine and pyrimidine analogues have been tested. Other such substances that have been used extensively are *5-fluorouracil* and *5-fluorodeoxyuridine,* which at relatively low concentrations interfere with DNA synthesis by inhibiting thymidine synthetase. At higher concentrations these substances also inhibit RNA synthesis and protein synthesis.

However, whereas such agents may act in this manner, it is possible that they also act in other ways. Unfortunately, many such analogues have too great an effect on cell division in germinative areas (e.g., on blood-forming cells of man) to be useful for treatment of cancer. *Mitomycin C* interferes with DNA synthesis without markedly affecting RNA synthesis or protein synthesis. *Actinomycin D* at concentrations which have no effect on DNA synthesis blocks DNAdependent RNA synthesis, probably mainly messenger RNA production (Kihlman, 1966). Both antibiotics retard cell division and have been used extensively in recent years as selective inhibitors of the synthesis of the two nucleic acids. Such poisons generally prevent cells from entering mitosis.

Antagonists to folic acid such as amethopterin and aminopterin affect a specific step in the separation of the chromosome between metaphase and anaphase. They have therefore been classed as *specific metaphase inhibitors*. These experiments demonstrate that one of the B vitamins controls a specific step in the mitoses of the cells (chick fibroblasts) upon which they were tested. However, according to Taylor cells are arrested by such poison in either the $G_1$ or the early S phase and do not progress unless thymidylie acid becomes available.

*Alkylating agents (e.g.,* those which replace an H atom of a compound with an alkyl group such as the methyl or ethyl) do not affect growth but prevent cell division. The nitrogen mustards (used as mutagens and in the treatment of cancer) and methane sulfonates (e.g., Myleran used in treatment of chronic myeloid leukemia) are such alkylating agents.

The exact manner in which inhibition of cell division occurs is unknown, but some of these alkylating agents cause chromosome breakage or failure of normal chromosome movement similar to the effects following x-radiation. This could presumably occur as a result of translocation of parts of two chromosomes, resulting in two kinetochores (chromosome constrictions for spindle fiber attachment) at two places on a single chromosome which is torn in half during anaphase. Abnormal disjunction could occur from failure of attachment of kinetochores. These agents also appear to affect the formation of spindle fibers, generally reducing the viscosity of the cytoplasm. Esterification of carboxyl groups of proteins and combinations with nucleic acids have also been demonstrated, but the precise nature of the reactions is unknown.

The antimitotic agents such as *colchicine, podophyllin* and the *urethans* inhibit the formation, or the breakdown of the mitotic apparatus. Colchicine apparently inhibits cell division by disorganising the mitotic spindle, without stopping growth or duplication of various organelles in the cell. Chromosomes duplicate themselves, but the spindle fibers that form are disoriented, resulting in polyploidy which, in some species, persists even after the poison is removed. It does not interfere with the making or breaking of disulfide bonds between proteins forming the spindle fibers, but seems to interfere with the secondary bonding in the formation of an oriented and symmetrical mitotic spindle.

The *sulfhydryl reactants (e.g.,* chloracetophenone), on the other hand, block cell division by interfering with the disulfide and sulfhydryl cycle which is necessary for the formation and dissolution of the spindle fibers. All the other processes continue up to the formation of the spindle.

Another interesting poison which has given considerable insight into mitosis is *mercaptoethanol* which, among other things, prevents the duplication of *mitotic centers* (centrioles) of the cell. Centrioles always appear in animal cells in pairs. Each member of a pair must divide before they can act as division centers for the cell. The precise way in which mercaptoethanol affects the pattern of cell division depends upon just when it is applied during the division cycle.

Exposure of the sea urchin eggs to this poison delays cell division to the extent that it delays duplication of the mitotic centers, which divide only after the cells are removed from the poison. In addition to this delay, cell division is made abnormal because, although the division of centrioles is inhibited by mercaptoethanol, the separation of the centrioles from one another is not affected.

Along with all the other processes preparing for cell division, the members of the pair of centrioles must separate and take their positions at the poles of the cell. If this poison is applied to an egg cell just before its first division when it already has two pairs of centrioles and the egg is kept in mercaptoethanol beyond the time for the first division, it will then prepare for the second division except that the required duplication of the centrioles will not occur. The two pairs of centrioles separate and single centrioles now occupy four coordinate radii of the egg. Released from mercaptoethanol after being washed in sea water, the centrioles first divide and then the egg divides into four cells at once. The data are of interest because they show that the poison specifically affects only the duplication of centrioles,-not their separation or movement in the cell.

Some antibiotics, such *as penicillin,* prevent cell division of bacteria by interfering with the incorporation of amino acids into the cell walls and causing some bacteria to develop as protoplasts-without cell walls. Such protoplasts cannot divide-apparently the cell wall is necessary for bacterial cell division. Since animal cells do not form cell walls they are unaffected by this action of penicillin.

A search is being made for naturally occurring substances with dual actions, on the one hand promoting growth, and on the other, retarding growth. Such substances are induced in animals after injection with active cancerous cells. Chemical analysis suggests keto-aldehydes as the compounds produced in response to such injections. Such compounds are being tested as possible regulators of cell division.

## THE INITIATION OF CELL DIVISION

The cell, as we have seen, grows to a certain stage and then divides. In some unknown way each unit (organelle?) in the cell presents a template upon which duplication occurs. When cell division is completed each daughter cell contains in itself a duplicate of all the varied structures present in the mother cell.

Cell division has been studied in a multitude of cells, and most cells, under normal conditions, are known to divide by mitosis. Chromosomes condense and appear on an equatorial plane. The daughter chromosomes, produced by splitting, separate from one another, and a duplicate set of chromosomes is moved toward each of the two poles. Finally the cytoplasm separates into two portions. The complete process may take no more than half an hour in some marine eggs. Mitosis is described in detail in textbooks of cytology. The interest here centers on the mechanism initiating cell division.

The trigger which sets off the cell division has been of interest for a long time and numerous suggestions have been made about the nature of this process. One possibility that suggests itself is the doubling in mass of the cell, since this might be the last growth requirement for a series of processes, each resulting in doubling of cell organelles.

Division then occurs because the ratio of cytoplasm to nucleus is upset. Some evidence points this way. As previously indicated, when an ameba has doubled its mass it divides, but not immediately. The division takes place after a lapse of time (rest period) during which something occurs which triggers the division of the cell. This time lapse before division is about 4 hours, or about one-sixth of the generation time of 24 hours for *Amoeba proteus*.

Also, when an ameba about to divide is shaken, it divides unequally, but each daughter ameba grows until it equals the mass of the mother ameba before it divides again. Moreover, amputation of a part of a growing ameba is followed by replacement of the mass removed before division occurs at all. In all these cases division takes place 4 hours after the ameba has doubled its mass. Growth before division in all cells is not alike however. *Tetrahymena* shows an increase in mass followed by a predivison "rest period" similar *to Amoeba*, but in *Escherichia coli B,* synthesis proceeds in exponential fashion.

Another suggestion for the trigger in cell division is the upsetting of the surface to volume ratio. However, if a piece of cytoplasm is removed at a "critical" time just before an ameba has doubled its mass, division will still occur, even though smaller offspring than normal are produced. This experiment might be taken to indicate that, by this time, division has been determined and follows even if the ameba is not as large as a typical mother cell about to divide, and that neither the ratio of nucleus to cytoplasm nor the ratio of surface to volume can in itself be the probable trigger for cell division. When cells are starved they may undergo division producing smaller daughters. However, such alterations in cell size are also produced by a number of other changes in environment.

Doubling the DNA content has been suggested as still another possibility. By the use of labeled precursors, it has been shown in most cells, however, that DNA is doubled in the interphase long before division actually occurs, in fact even before the formation of the extra pairs of chromosomes. Doubling of DNA content is therefore not the likely trigger. As the nucleolus re-forms in the "resting" nucleus, intense protein synthesis appears to occur just outside the nuclear membrane. The interphase is apparently not a period of rest for the cell but rather a period during which the necessary cell constituents are manufactured by it.

The possibility that the nucleolus has a vital role in initiating cell division is suggested by experiments in which microbeam irradiation

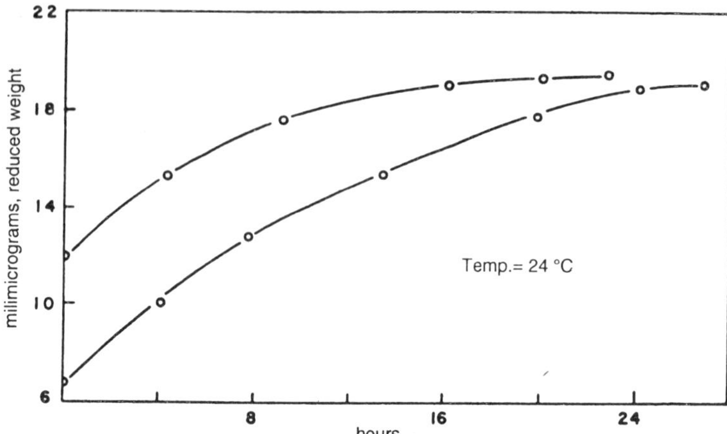


*Figure 9.9: Relation between the weight of Amoeba proteus at "birth" and the weight at maturity. Curves represent the growth of sister cells following unequal division of their mother cell. The last point on each curve gives the weight at the time of division.*

of a nucleolus (neuroplast of a grasshopper) at certain critical times (telophase to middle prophase) permanently stops mitosis in the cell, while comparable irradiation of the surrounding areas of the cell does not.

It was also thought possible that the chromosomes or other organelles produce a specific substance which induces the division of a cell. One team of workers found that kinetin (6-furfurylaminopurine) serves this function in some plant cells in which division is otherwise infrequent. No convincing evidence has accumulated to show this to be a general phenomenon. In fact, it has been shown that nuclear structures need not be present for the cytoplasm of the sea urchin egg to divide, since non-nucleated egg fragments can be excited to divide by parthenogenetic agents. However, such cells develop abnormally and eventually die since they have no nuclei.

According to Heilbrunn, cell division is initiated by a coagulative process similar to the formation of a clot in blood, and any stimulus which induces clot formation is likely to initiate formation of spindles and cell division. It has been shown that agents which inhibit clot formation, e.g., heparin, interfere with the division of normal fertilised eggs. Agents which release calcium, which is also required for clotting blood, facilitate cell division, whereas agents which bind calcium, such as oxalate or citrate, inhibit cell division.

Whatever the immediate cause of cell division, it is clear that there are apparently many factors that contribute toward a favourable state for division of the cell and that preparations must be completed along many parallel lines before cell division can occur.

## THE NATURE AND FORMATION OF THE MITOTIC APPARATUS

That the division spindles are gels has long been known. This was demonstrated by micromanipulative studies and by experiments in which intact mitotic figures were isolated from dividing eggs by the use of mild detergents which solubilise the rest of the cell. The mitotic apparatus has also been isolated from dividing eggs by changes in pH and from dividing eggs from which the membrane and the hyaline outer cytoplasm has been removed by immersion in Versene (ethylenediamine tetraacetic acid), dextrose and dithioglycol solutions. The mitotic spindle is made up of about 3 to 5 per cent ribonucleic acid (RNA), and the remainder appears to be largely one type of protein.

Electrophoretic diagrams indicate two peaks, one of which is the protein, and the other its conjugate with RNA; both act as antigens. Electron microscope studies of fixed cells demonstrate that the spindle gel consists of definitely organised and oriented microtubules of protein, the centers of orientation being the centrioles. When three pairs of centrioles are present, as in polyspermic eggs, the orientation is toward three poles. Presumably, something in or from the centrioles leads to the development of such oriented tubules, although no evidence for its origin or transport has yet been obtained.

On the basis of chemical evidence, it was suggested that the linkages between protein molecules in the microtubules of the mitotic spindle are disulfide (S-S) bonds. In the preparatory phase the *intramolecular* disulfide bonds of protein molecules are presumably reduced by glutathione. This is considered to take place before the metaphase, at a time when the glutathione content of the egg is known to be declining. When the mitotic apparatus is fully formed and growing, the disulfide bonds are supposedly restored, but now as *intermolecular* disulfide bonds, binding the protein molecules into tubules and liberating the glutathione.

The concentration of the glutathione in the egg cell was found to increase at this time. Postulation of intermolecular disulfide bonds rests on the assumption that the probability of forming a bond between sulfur atoms on two protein molecules would be greater than the probability of re-forming the original intramolecular disulfide bonds. In essence, then, the sulfhydryl cycle has converted intramolecular disulfide bonds into intermolecular disulfide bonds, the glutathione acting essentially like a catalyst, being recovered unaltered at the end of the cycle.

Although the cyclic variation in SH bonds present in dividing cells has been confirmed independently, there is little evidence to back the specific hypothesis just outlined.

If orientation and organisation of the spindle tubules are a prime necessity for cell division, then any agent which interferes with the spindle should interfere with cell division, even if this agent permits gelation. This is exactly what happens when a cell is poisoned with colchicine. A gel appears, but the secondary bonds between protein molecules fail to develop, the gel is disoriented and separation of the doubled chromosomes fails to occur ("*flabby anaphase*"). Because of this, as we have already seen, polyploid cells are produced by the use of colchicine. When such cells are washed free of colchicine, they later divide, but the polyploid condition is maintained.

When Hoffmann-Berling placed cells about to divide (i.e., showing spindles) in glycerin solutions, the division process ceased. When they were later placed in a balanced salt solution, the spindles remained, but division still failed to occur. It is perhaps possible that some material necessary for cell division had leached into the glycerin solution. When ATP was applied to such mitotic "models" (glycerin-extracted dividing cells) the spindle tubules contracted and the chromosomes moved toward the poles.

Furthermore, an enzyme for decomposing ATP is present in the spindle. On the other hand, ATP produced no movement when applied to mitotic spindles isolated from sea urchin eggs by mild detergents and subsequently alcohol hardened. Hoffmann-Berling points out that hardening the spindles denatures the proteins of the tubules, making them incapable of contraction in response to ATP, even though hardening permits their separation and purification from other cellular constituents.

It is an oversimplification to consider only the contraction of the spindle tubules. Careful analysis of the movements during mitosis indicates that only the early part of mitosis (to anaphase) occurs by contraction. This is indicated by the broadening of the spindle. Evidence suggests that in anaphase the two sets of chromosomes in the cell are actually pushed apart by the tubules between the poles. For example, when the spindles of grasshopper neuroblasts are immersed in hypertonic solutions which reduce the volume of the cell and thus crowd the mitotic spindle, the chromosomes are crumpled and spiraled during the anaphase and pushed against the ends of the cell.

The active elongation of the spindle poses the same problems as elongation and relaxation of muscle fibers, both active processes. It has been shown that low concentrations of ATP applied to the mitotic models cause contraction; higher concentrations, elongation. On the other hand, the spindle fibers appear to consume themselves as contraction occurs, unlike a muscle which maintains its mass in spite of shortening. A single contraction of the spindle tubules, however achieved, suffices to separate the duplicated chromosomes from one another. It would almost appear as if the chromosomes, as they move toward the poles, release some material which dissolves the spindle tubules.

The fact that glycerinextracted models of dividing cells contract when placed in ATP, and that the chromosomes on the metaphase plate separate, could conceivably be explained as the result of contraction of the entire cell in such a way as to actually simulate mitosis rather than by contraction of the spindle tubules. The problem of chromosome movement should not be considered solved, despite the appeal of the contractile tubule hypothesis.

## FURROW FORMATION AND CLEAVAGE (CYTOKINESIS)

The formation of a furrow in a dividing egg or other animal cell appears to be accompanied by increased viscosity in the area of the furrow. High pressures, which liquefy protein gels, also prevent the formation of cleavage furrows and cause such dividing cells to fuse again. The furrow disappears with even a slight pressure applied to both sides with a microneedle, but cell division resumes when the pressure is released. An egg ruptured at the time of furrow formation disintegrates, but a part of the cortex and the furrow remain intact, indicating that the furrow is apparently a rigid part of the cell. Furthermore, it has been noticed that as the cortex forms in the cleaving eggs, a slight separation between the cortical layers of the two blastomeres (the twocelled stage) is maintained for a time, indicating that some change has occurred which prevents coalescence of the two layers.

An attempt has been made to explain the formation of a furrow in a dividing cell by means of the *contractile ring theory* (or *cortical gel contraction theory*). According to this theory the Belated ring in the cortex of the dividing cell is considered to contract like the nonmotile portion of an ameba, and it might be expected, therefore, to decrease the surface area. However, measurement in some cells showed that before division the surface increased by about 26 per cent.

The increase in surface area accompanying cell division does not invalidate contraction theories of cell division. It is essential that the cell elongate before it cleaves, and at this time the surface area

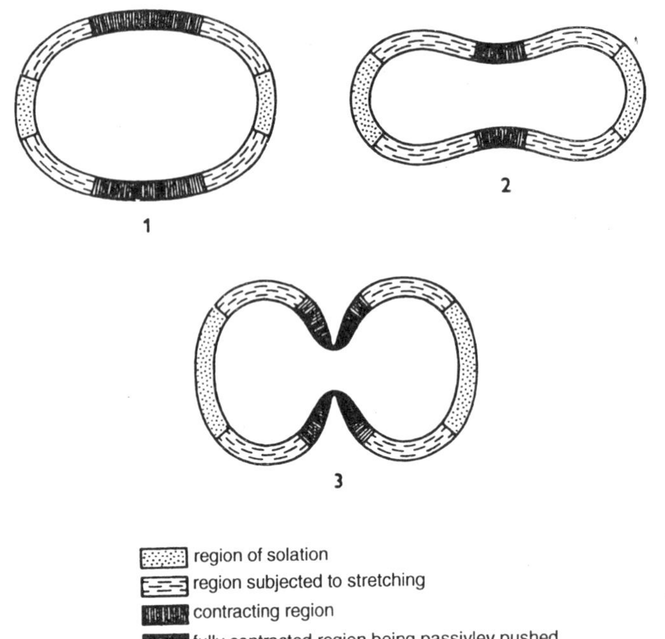


*Figure 9.10: Diagram of the cortical gel contraction theory of Marsland and Landau. 1, Early telophase; 2, early furrow; 3, late furrow.*

increases. But elongation is part and parcel of any contractile system, as already seen in muscle fibers. When relaxation of muscle cells is better understood it may also be possible to understand cell division better. In modified form the contractile furrow theory, illustrated in Figure elsewhere in this chapter, allows for a passive increase in surface at the poles as a result of active contraction of the furrow region.

The *expanding surface theory* on the other hand, suggests that a nuclear substance is liberated (probably from chromosomes) which causes expansion of the cellular membrane at the poles. As the polar area expands, the equator contracts, leading to division. This theory is supported by two lines of evidence. First, movement of material on the surface of cells, consistent with such a suggestion, has been observed.

Second, expansion at the poles should weaken the membrane there. The fact that cytolysis of an egg occurs more readily during division than during interphase suggests some change in the cell membrane at the poles. The equatorial part persists after cytolysis, apparently being

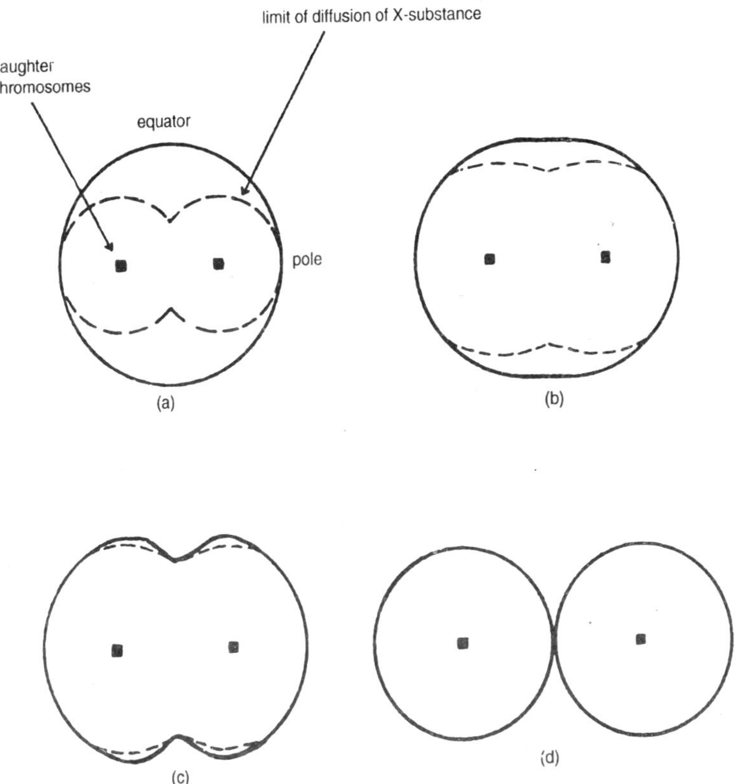


*Figure 9.11: Diagram to illustrate the expanding membrane theory of cleavage of Mitchison and Swann. The X-substance is the hypothetical substance which presumably comes from the chromosomes, and which leads to an increase in polar surface area upon contact with the poles.*

more rigid. However, even cells without nuclei or chromosomes cleave, indicating that the excitant may have origins other than the nucleus. Another form of this theory postulates expansion of surface area as a result of growth. It is interesting that the resting potential of a cleaving starfish egg does not change.

The *spindle elongation theory* assigns a decisive role to the spindle and asters in cell division. The evidence comes from observation of movement, with respect to one another, of kaolin particles attached to the surface of an egg membrane. The basic observation is that the elongation of the cell at anaphase is accompanied by a shrinkage at the equator, the two kaolin particles on either side of the equator coming closer together while there is a corresponding stretching at the poles. Since the spindle and asters appear to be rigid structures, as tested with

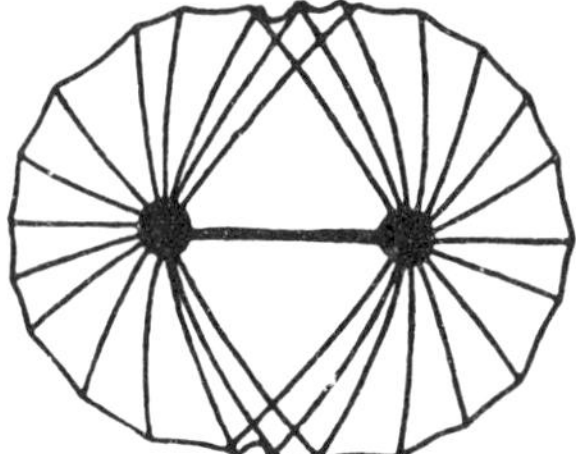
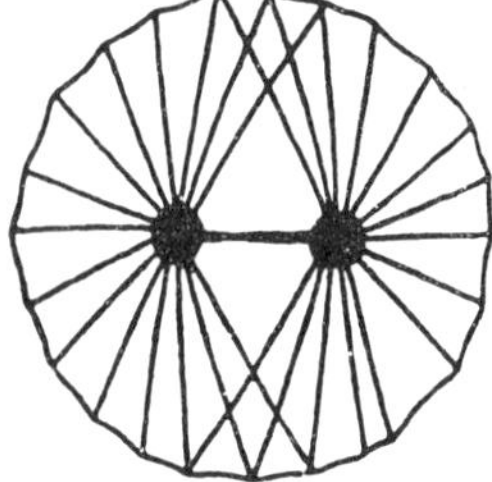

*Figure 9.12: Model illustrating Dan's spindle elongation theory. Note especially the effect of the spindle elongation upon the astral ray attachments at the surface of the cell. Spindle elongation results in a pull at the furrow region and furrow formation.*

a micromanipulator, the driving force is believed to be the elongation of the spindle tubules which pushes the centers apart, bringing the astral rays attached to the equator close together.

Because a nucleus-free half of a sea urchin egg will cleave when artificially stimulated to divide, the *astral relaxation theory,* based only on changes produced by the asters, has been proposed to account for cleavage. Surface tension and elasticity measurements show that an egg surface is under uniform tension before cleavage begins. When the asters reach the poles of the egg it is thought that they produce a change (the nature of which is unknown), and the surface tension at the poles is lowered. This permits the furrow region, which maintains its surface tension, to contract while the polar regions expand.

Experiments of Hiramoto, however, in which the mitotic apparatus was either sucked out or displaced by injection of mineral oil into the egg, indicate that even in the absence of spindle tubule attachments (asters), cleavage furrows are formed in the same place as they are when mitotic apparatus is intact and in place. Also, a piece of cortex of an amphibian egg develops a furrow even in the absence of a nucleus and a spindle. Stiffness of the cortex of the sea urchin egg increases before cleavage furrows form and varies cyclically during the division interval. The contracting power of the equatorial ring of the sea urchin egg reaches a maximum at the middle of the cleavage cycle.

Work with eggs under pressure and subjected to various inhibitory reagents suggests that the ATP system accounts for the energy of whatever movements occur during cleavage, since addition of ATP often relieves the effect of an inhibitor. The energy required for division of an egg has been calculated to be about three times that normally available from respiration. Since the rate of respiration rises only a few per cent during cleavage, it would appear that the energy for cleavage must come from a specific source. It is suggested that compounds with high energy phosphate bonds, formed by the cell in preparation for each successive cleavage, might well serve this purpose.

It is apparent that all of these theories have much in common and that they are developed primarily from data on the marine egg cell. They do not apply to such phenomena as multiple fission, which occurs in some plant and animal cells, nor do they take into consideration the division of plant cells with cell wallswhere the laying down of a new wall is of prime importance for cleavage. In bacteria and yeast, as was seen earlier, removal of the cell wall prevents cell division even though the cells continue to increase in size. We are still without a generalized theory of cytokinesis.

# INDEX

## A

## B



## C



## D

## G



## H



## I

## Q



## R



## S

## T